A Practical Guide to XLIFF 2.0

Bryan Schnabel

JoAnn T. Hackos

Rodolfo M. Raya

A Practical Guide to XLIFF 2.0

Credits

Cover image:	Courtesy of SimpleIcons, openclipart.org, public domain.
World image Chapter 7:	Courtesy of neocreo, openclipart.org, public domain.
Fender Stratocaster Guitar Chapter 7:	Courtesy of Kamil Stepinski, openclipart.org, public domain.
Fender Telecaster Guitar Chapter 7:	Copyright © 2010 Ethan Prater, CC BY 2.0.

Disclaimer

The information in this book is provided on an "as is" basis, without warranty. While every effort has been taken by the authors and XML Press in the preparation of this book, the authors and XML Press shall have neither liability nor responsibility to any person or entity with respect to any loss or damages arising from the information contained herein.

This book contains links to third-party web sites that are not under the control of the authors or XML Press. The authors and XML Press are not responsible for the content of any linked site. Inclusion of a link in this book does not imply that the authors or XML Press endorse or accept any responsibility for the content of that third-party site.

Trademarks

XML Press and the XML Press logo are trademarks of XML Press.

All terms mentioned in this book that are known to be trademarks or service marks have been capitalized as appropriate. Use of a term in this book should not be regarded as affecting the validity of any trademark or service mark.

XML Press
Laguna Hills, California
http://xmlpress.net

First Edition
978-1-937434-14-4 (print)
978-1-937434-15-1 (ebook)

Table of Contents

List of Examples

Preface

About this book

A Practical Guide to XLIFF 2.0 introduces the OASIS XLIFF standard and gives you, as a potential user of XLIFF, the information you need to create XLIFF files and workflows. It reviews how to do the following:

- exchange XLIFF files
- consume translated XLIFF files
- use best practices in the XLIFF translation workflow
- validate XLIFF files
- follow the processing requirements defined in the XLIFF standard
- extend XLIFF in compliance with the standard
- add modules and core features to future versions of XLIFF

The audience for this book

This book is written for the following audiences:

- Localization coordinators at large companies who currently use out-dated methods to manage translations and now must manage XML text, software strings, RTF, Graphics, DITA, HTML, SVG, and more.
- New localization coordinators who need guidance and better methods to manage their translation projects.
- Technical writers who are responsible for managing translations without the assistance of localization coordinators.
- Those who prefer to handle translation using open standards rather than proprietary systems.
- Localization Services Providers (LSP) who prefer to handle DITA and other content using open standards rather than proprietary systems.
- Content Management System (CMS) vendors and other software developers who want to incorporate XLIFF into their publishing pipeline.
- Consultants who want to incorporate XLIFF into their customers' publishing pipelines.

Typographical conventions for this book

This book uses the following typographical conventions:

`Constant-width font`	XML `<elements>`, XML `attributes`, and code examples.
Italic font	Glossary terms and emphasis.

The authors

Bryan Schnabel, Rodolfo M. Raya, and JoAnn Hackos are active in the XLIFF community.

Bryan Schnabel

Bryan is chair of the OASIS XLIFF Technical Committee. He has written several XLIFF and DITA software programs. He lives in the United States in Sherwood, Oregon. When he's not programming and working with XLIFF, he is usually creating music. At the time of this writing his first solo album has been released, and the second is on its way.

Rodolfo M. Raya

Rodolfo was formerly Secretary of the OASIS XLIFF Technical Committee. He is CTO (Chief Technical Officer) at Maxprograms,[1] where he develops cross-platform translation/localization and content publishing tools using XML and Java technology. Rodolfo lives in Montevideo, Uruguay, and dedicates his free time to sailing with family and friends.

JoAnn Hackos

JoAnn is the founder and president of Comtech Services, Inc.,[2] an information strategy consultancy headquartered in Denver, Colorado. JoAnn chairs the OASIS DITA Adoption Technical Committee and was one of the founders of the DITA standard. She is an avid birder and travels around the world seeking new species and new experiences in nature.

---

[1] http://maxprograms.com
[2] http://www.comtech-serv.com/

Acknowledgments

Bryan Schnabel, Rodolfo Raya, and JoAnn Hackos would like to make the following acknowledgments.

Bryan

Thanks to my wife, Patricia, for her steadfast support throughout this project, Rodolfo, for his field expertise and for keeping the scope of the book true, and JoAnn for her real-world understanding of how to represent the readers' point of view and for teaching me how each word (said or unsaid) matters in this book. I would also like to thank Yves Savourel and David Filip for reviewing the book, and Mt. Hood Meadows for providing a place to ski and unwind.

Rodolfo

Many thanks to Bryan for his patience during the writing experience and to JoAnn for her continuous support.

JoAnn

I'd like to thank Bryan Schnabel for all of the great work he has done to develop the XLIFF standard and to create the core content of this book. He is very much the key contributor. I'd also like to thank Rodolfo Raya for educating me about the best practices involved with translation and translation management.

Getting Started

A Practical Guide to XLIFF 2.0 is structured to accommodate the diverse audiences that will use XLIFF and need to understand how to manage it effectively.

In Part I, "Getting Started," we orient you to the OASIS XLIFF standard, describing how it works and how it fits into the translation process. We also describe its architecture, the difference between the XLIFF core and its modules, and the XLIFF conformance requirements. We recommend that everyone start by reading this section.

Beyond Part I, we recommend you focus on different parts of the book based on your role. Here are some typical roles and our recommendation for each.

Translation Architects: A translation architect is responsible for translations on behalf of a company, across departments, deliverables, and formats. This person owns the success or failure of the translation strategy. In this role, you are expected to know all aspects of the workflow, tools, schedules, roles, financial impacts, and technology stack.

If you are or wish to become a translation architect, you should read all four parts of this book.

Localization Coordinators: A localization coordinator is responsible for the translation of a specific workflow. This person coordinates the workflow with the Localization Service Provider (LSP), the content providers, and the publication stream and, in some cases, manages the budget for translations.

If you are or wish to become a localization coordinator, you should read all of Part I and the chapters that discuss the specific tasks for your role in Part II, "Applied XLIFF." For example, if you are the localization coordinator for a web team, you should read Chapter 5, *Translating Websites*. In Part III, "XLIFF Core," you should at least read the introductory material about

XLIFF core features and functions. And in Part IV, "XLIFF Modules," you should at least read the introductory material.

Technical Writers: Technical writers are often responsible for sending DITA and other XML-based content to translators without the assistance of localization coordinators.

If you are a technical writer, you should read all of Part I, and you should read the chapters that discuss the specific tasks for your role in Part II. For example, if you are a technical writer in a DITA publishing team, you should read Chapter 4, *Translating DITA*. In Part III, you should at least read the introductory material around XLIFF Core Features and Functions, and in Part IV, you should at least read the introductory material.

Localization Services Providers: Localization Service Providers (LSP) have a unique relationship with XLIFF. Computer-Aided Translation (CAT) tools generally use XLIFF as their native file format. And, more and more, informed translation customers send XLIFF to their LSPs instead of source content.

If you are a Localization Services Provider, you should read all of Part I, and you should read the specific tasks your customer is involved with in Part II. For example, if your customer is translating graphics, you should read Chapter 7, *Translating Graphics*. You should read all of Part III and Part IV.

CMS and other XML software developers: As a developer of content management systems (CMS) and XML software, you will need to know how to incorporate XLIFF into your publishing pipeline.

If you are a developer, you should read all of Part I, and you should read the the chapters that discuss the specific tasks your product handles in Part II. For example, if your platform is translating XML, you should read Chapter 3, *Translating XML*. You should read all of Part III, and in Part IV, you should read about any module that will enhance your translation workflow.

Consulting Groups: Consulting groups often want to incorporate XLIFF into the publishing pipelines they design for customers. If you are or wish to become a consultant, you should read the entire book.

CHAPTER I

Introducing XLIFF

The need to provide information to people who speak a different language is an age-old challenge. The evolution of the translation process over the years looks something like this:

- In the beginning, translations were done by brute force.
- Over time, successful methods became repeatable processes that could be documented and shared from one group to another.
- As momentum grew, these processes became ad-hoc standards.
- The game-changer, the watershed moment, occurred when we recognized that ad-hoc standards must evolve into open standards.

Once XLIFF (XML Localisation Interchange File Format) and its fellow translation standards became open standards, tools, translators, content management systems, and translation customers could exchange translation workflows in a predictable way. And to sweeten the deal, the introduction of open standards for content, such as DITA, SVG, DocBook, and HTML, made it possible to automate the translation workflow using standard tools.

What is XLIFF?

XLIFF is an XML standard developed under the auspices of OASIS Open, a "non-profit consortium that drives the development, convergence and adoption of open standards for the global information society."[1] OASIS XLIFF 2.0 is the current version of the standard.

XLIFF was developed to facilitate the exchange of content during localization and reduce the number of document formats that localization companies receive from information developers. XLIFF enables information developers – whether they create product documentation, training materials, or entire websites – to reduce their translation and localization costs. XLIFF has been adopted by the translation industry as an exchange format for manuals and text-centric documents.

While translation focuses on expressing the meaning of a piece of text in a different language, localization goes beyond translation by adapting the translated content to the needs of the intended

---

[1] http://www.oasis-open.org

audience. While the translation process focuses on the text, localization also considers colors, shapes, date and time use, and other details specific to the country or place where a product is to be used. The XLIFF format includes attributes that support these localization tasks.

The translation process: extract and merge

Translating with XLIFF begins with the *Extract & Merge* method. A person or process extracts translatable text from a *source file* and stores it in an XLIFF document. The source file's structure is preserved in an auxiliary file called a *skeleton*. After translation, a person or process merges the translated XLIFF file and the skeleton back into the source file format. The new file is the translated *target file*.

Figure 1.1 shows the minimal translation workflow in XLIFF using the extract and merge method.

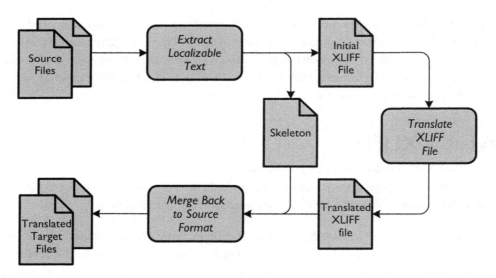

Figure 1.1 – Extract and merge workflow

1. Extract the localizable text into suitable segments and place it in XLIFF `<source>` elements within `<unit>` and `<segment>` elements (see Example 1.1).

   Example 1.1 – Text extracted into segments

   ```
   <unit id="title-2">
     <segment>
       <source>Birds in Oregon</source>
     </segment>
   </unit>
   ```

2. Capture metadata and structure information from the source file. You have two options:

   a. Store the metadata and structure information in a skeleton file. This method is called the *minimalist method*. The skeleton file can be embedded in the XLIFF file or stored as a separate document. Refer to the section titled "The minimalist method" (p. 70) for more information about the minimalist method.

   b. Capture the metadata and structure information in `<group>` elements. This method is called the *maximalist method*. Refer to the section titled "The maximalist method" (p. 73) for more information about the maximalist method.

3. Translate and save the text in the XLIFF file using a translation tool. The file will now be bilingual, with `<source>` elements containing the original text in the source language and `<target>` elements containing the corresponding translation in the target language (see Example 1.2).

   Example 1.2 – Text with `<source>` and `<target>` elements

   ```
   <unit id="title-2">
     <segment>
       <source>Birds in Oregon</source>
       <target>Pájaros en Oregon</target>
     </segment>
   </unit>
   ```

4. Using the metadata and structure information, convert the XLIFF file back into its original format, but with the translated text in place of the original.

Several commercial and open source tools exist that will perform the extract and merge operations.

The translation process: first time

Extract and merge is just part of the process. If you will be translating a set of documents multiple times or you have similar documents that need to be translated, you must add steps to the initial translation process, including creating translation memory and bilingual glossaries from the translated XLIFF.

To populate or update translation memory, you can convert XLIFF files to the TMX (Translation Memory eXchange) format using XSL transformations or export to the TMX format using a translation tool. You can extract a bilingual glossary from translated XLIFF or TMX files using a translation tool and store the results in a glossary database.

Figure 1.2 shows a comprehensive workflow diagram for the initial translation process for source documents in a content repository, including steps to update translation memory and a glossary. Optional steps are shown in grey.

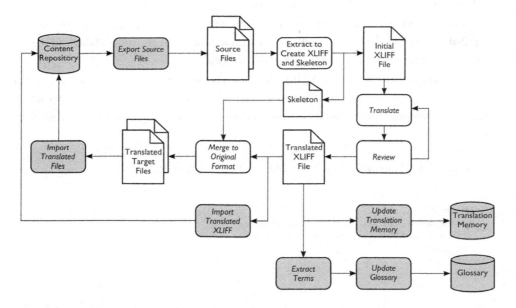

Figure 1.2 – First-time translation workflow

The translation process: maintenance mode

When you are ready to translate updated content, you expand the first-time workflow to include additional steps that facilitate translation and reduce overall costs.

The expanded set of steps is as follows:

1. Separate translatable text from markup, storing the translatable text in an XLIFF file and the formatting data in a skeleton file.
2. If you have not done so while creating the XLIFF file, segment large text fragments.
3. Compare the new XLIFF file with the previously translated one and recover exact matches – segments that have remained unchanged between versions.
4. Merge the translations for any exact matches into the new XLIFF file and mark the exact matches as untranslatable.
5. Use the translation memory created in the initial workflow to locate matches for untranslated segments and create a partially translated XLIFF file.
6. Use translation memory and the glossaries to generate an *example-based machine translation* for each untranslated segment.
7. Send the resulting XLIFF file to an LSP for translation and review.
8. Merge the translated XLIFF file with the skeleton to generate a translated document.
9. Update the translation memory and glossary with the results of the translation.
10. Save the translated source and translated XLIFF in the content management system (CMS).

As noted above, the new translations can be used to populate your translation memory and bilingual glossaries.

Figure 1.3 shows the complete workflow for the document maintenance cycle. The grey boxes are the new parts of the workflow.

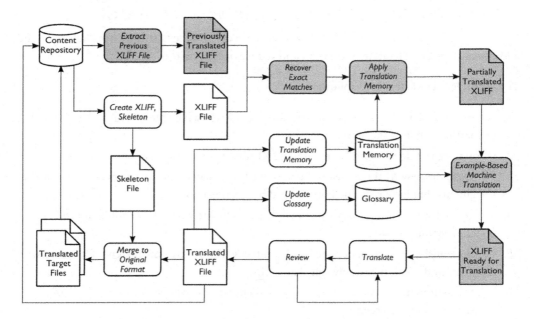

Figure 1.3 – Maintenance mode translation workflow

XLIFF 2.0

The current version of XLIFF is 2.0, released in 2014. Each published version of the XLIFF specification links to the previous version. You can follow those links to browse the history of XLIFF releases.

- XLIFF 2.0 Committee Specification 01 was approved on 31 March 2014. This version is available at http://docs.oasis-open.org/xliff/xliff-core/v2.0/cs01/xliff-core-v2.0-cs01.html.
- The first official release of XLIFF 2.0 as an OASIS Standard was approved on 5 August 2014 and is available at http://docs.oasis-open.org/xliff/xliff-core/v2.0/xliff-core-v2.0.html.

XLIFF 1.x

During the XLIFF 1.x era, the latest approved XLIFF specification was always available (and can still be found) at http://docs.oasis-open.org/xliff/xliff-core/xliff-core.html.

- The initial draft for XLIFF 1.0 contributed to OASIS was posted on 27 April 2001. It is available at http://www.oasis-open.org/committees/xliff/documents/contribution-xliff-20010530.htm.

- The XLIFF 1.0 Committee Specification was last published on 15 April 2002. It is available at http://www.oasis-open.org/committees/xliff/documents/xliff-20020415.htm.

- The XLIFF 1.1 Committee Specification was last published on 31 October 2003. It is available at http://www.oasis-open.org/committees/xliff/documents/cs-xliff-core-1.1-20031031.htm.

- XLIFF 1.2 was the first XLIFF version to achieve fill OASIS standard status. It is available at http://docs.oasis-open.org/xliff/v1.2/os/xliff-core.html. It became an OASIS Standard in 1 February 2008.

CHAPTER 2
Introducing XLIFF 2.0

XLIFF 2.0 introduces radical changes. The standard was redesigned from the ground up to offer better support to the localization and translation industry. The changes include the schema, specification, and conformance requirements.

XLIFF 2.0 was written to be extensible. It provides a basic set of elements and attributes in the XLIFF Core. The Core supports the preparation of basic XLIFF files containing only the material that needs to be translated and supports the generation of a translated version of the original document.

Any tool that claims to support XLIFF should handle XLIFF files containing only core elements and attributes.

Some features present in XLIFF 1.x, such as storage of suggested translations, are considered advanced features in XLIFF 2.0. These features are now defined in separate XML schemas and distributed as XLIFF 2.0 optional modules.

XLIFF is an XML vocabulary. Its specification includes XML schemas against which the XML must be valid. You will find a link to download the XML schemas and XML catalog at the beginning of the specification.[1]

In addition to the XML schemas that the XLIFF file must validate against, XLIFF 2.0 includes a robust set of documented constraints and processing requirements.

OASIS XLIFF Technical Committee provides a set of test suites.[2] In addition, you can find a variety of utilities to check constraints and processing requirements. At the time of publication, the availability of utilities was in flux. An internet search for "XLIFF utilities" will help you locate utilities as they become available.

---

[1] The XLIFF 2.0 schemas and catalog can be found at: http://docs.oasis-open.org/xliff/xliff-core/v2.0/os/schemas/

[2] https://tools.oasis-open.org/version-control/svn/xliff/trunk/xliff-20/test-suite/

XLIFF 2.0 Core

The XLIFF Core supports the minimum set of tasks required to do a functional translation from a single source language to a single target language. Any XML processor that claims to implement XLIFF 2.0 must support the XLIFF 2.0 Core. The core includes markup to support the following capabilities:

- Translation metadata management
- White space processing
- Source format and hierarchy management
- Extract and merge operations
- Grouping and segmenting
- Source text markup
- Translated text management
- Markup to identify which text can and cannot be translated
- Sub-flows
- Annotations
- Bidirectional text
- Fragment identification

XLIFF 2.0 Modules

XLIFF 2.0 defines optional modules that extend the XLIFF Core. XLIFF processors that support a given module must produce files that validate against the XLIFF 2.0 schemas and comply with that module's constraints. XLIFF 2.0 includes the following modules:

- Translation Candidates module
- Glossary module
- Format Style module
- Metadata module
- Resource Data module
- Change Tracking module
- Size and Length Restriction module
- Validation module

XLIFF 2.0 conformance

One of the shortcomings of XLIFF 1.2 was the lack of requirements for ensuring conformance, which led to interoperability issues between tools. XLIFF 2.0 strictly prescribes conformance. An XLIFF 2.0 file must be valid according to the XLIFF XML schemas, and it must also comply with the constraints and processing requirements documented in the XLIFF 2.0 specification. Any person, tool, or process that interacts with XLIFF 2.0 Core must adhere to the conformance requirements.

The XLIFF specification prescribes conformance for a variety of *agents*. The specification states that an agent is "any application or tool that generates (creates), reads, edits, writes, processes, stores, renders or otherwise handles XLIFF Documents."

Agents must read, generate, or process only valid XLIFF files, following the XLIFF 2.0 XML schemas, and they must comply with all applicable constraints and processing requirements. An agent must enforce the schemas and constraints both on input and output and, in particular, must reject invalid input, even if it might be able to recover. Depending on the part of the XLIFF process an agent supports, there are various requirements for conformance. Here are some of the agents and their requirements:

Writer or Writer Agent: Any agent that creates, generates, or outputs XLIFF 2.0 files. These files may be created for extractor, modifier, or enricher agents.

Extractor or Extractor Agent: Any agent that extracts localizable content from a native format and prepares it for translation into the target language, creating a valid and compliant XLIFF file.

Enricher or Enricher Agent: Any agent that incorporates module- or extension-based metadata and resources with extracted XLIFF content. An enricher agent expects valid and compliant XLIFF 2.0 input, and must generate valid and compliant output.

Modifier or Modifier Agent: Any agent that changes XLIFF structural or inline elements received from another agent. The result of the modification must be valid and compliant.

Merger or Merger Agent: Any agent that takes translated XLIFF 2.0 files and the original untranslated files in their native format and replaces the untranslated source language content with the localized content from the XLIFF 2.0 files. The resulting translated target files are native-format files with localized content in place of the source content.

Despite all the effort put into defining how XLIFF files should be prepared and processed, there are still places where things can go wrong:

- A modifier agent may replace the original inline elements written by an extractor agent with its own versions, creating potential problems for a merger agent.
- New inline elements may be added by a modifier agent. Although the merger agent is allowed to ignore those extra elements, the effects intended by the translator may be lost.
- Custom extensions used by extractor or enricher agents may not be understood by all modifier agents and, thus, may not be updated by a merger agent during the translation process.
- XLIFF XML schemas are public, but custom extensions defined in proprietary schemas are not required to be public or documented. This situation caused the most important interoperability problems in XLIFF 1.2, and that situation persists in 2.0.

Applied XLIFF

In Part 2, we present a series of tutorials that explain how to transform generic XML, DITA topics and maps, a website collection of HTML files, office documents, software strings, and other formats into and out of XLIFF. It ends with a discussion of how best to manage the translation process and avoid common pitfalls.

If you are a newcomer to XLIFF, we recommend that you review the step-by-step examples. In some cases, you will find references to functions that have not been explained, yet. We will cover those features in Part III and Part IV. However, this section will give you an overview of how XLIFF supports translation and localization.

CHAPTER 3
Translating XML

XML stands for eXtensible Markup Language, an open standard published and maintained by the World Wide Web Consortium (W3C).[1] The XML specification provides rules for creating custom vocabularies that follow a standard format.

XML is the key building block of many open standards, including XHTML, DocBook, DITA, OOXML (Office Open XML, the format used by Microsoft Office), Open Document Format for Office Applications (ODF, the format used by OpenOffice.org and derivatives), and XLIFF.

To translate XML content, you start with the extract and merge method described in the section titled "The translation process: extract and merge" (p. 4). The first step uses a specialized filter to separate translatable portions of an XML document from markup. Translatable portions are stored in an XLIFF file, and markup is stored in a skeleton file. After the XLIFF file has been translated, the translated XLIFF file and the skeleton file are merged to generate a translated instance of the source XML document.

XML documents may contain translatable text in these places:

- Plain text (PCDATA) inside an element
- Attribute values
- CDATA sections (content that processors interpret as character data, not markup)
- Comments

 Placing translatable text in CDATA sections or comments is not a best practice. However, since it happens, the XML workflow must account for this possibility.

There are two kinds of filters for converting XML files to XLIFF:

- **Generic**: filters that extract PCDATA from all XML elements
- **Vocabulary aware**: filters that selectively extract translatable content

---

[1] http://www.w3.org

Vocabulary-aware filters use configuration files that allow you to define which content should be translated and which shouldn't. A configuration file might include the following:

- Lists of translatable elements and attributes
- Catalogs for resolving XML schemas, DTDs, and entity modules
- Information about the content of CDATA sections

Because XML comments are usually not displayed by XML consumer applications, the text they contain is normally not translated. Nevertheless, if you are processing an XML document with comments, check with the author to determine whether comments should be translated.

Translatable content sometimes occurs in places that traditional XML mechanisms cannot handle. For example, CDATA sections are sometimes used as a work-around to add translatable content in areas that do not normally allow text. In such cases, you may need to use a special filter for CDATA content in addition to your usual filter.

In this section, we perform the following actions:

1. Extract localizable text in segments into XLIFF `<source>` elements.
2. Capture the metadata and structure information. See the section titled "The minimalist method" (p. 70) to use the minimalist method. See the section titled "The maximalist method" (p. 73) to use the maximalist method.
3. Translate and save the text in the XLIFF file using a standard translation tool. The file will now be bilingual, with `<source>` elements in the source language and `<target>` elements in the target language.
4. Using the metadata and structure information from the skeleton file, convert the XLIFF file back into its original XML format.

Consider the generic XML file in Example 3.1 as the source.

Example 3.1 – Generic XML file to be translated

```
<document status="draft" id="d1">
  <section id="s1">
    <title>Birds in Oregon</title>
    <paragraph>Oregon is a mostly temperate state. There are many
    different kinds of birds that thrive there.
    </paragraph>
    <section id="s1a">
      <title>High Altitude Birds</title>
      <paragraph>Birds that thrive in the high altitude include
      the White-tailed Ptarmigan, Sharp-tailed Grouse,
      Yellow-billed Loon, Cattle Egret, Gyrfalcon,
      Snowy Owl, Yellow-billed Cuckoo, and Boreal Owl.
      </paragraph>
    </section>
    <section id="s1b">
      <title>Ocean Birds</title>
      <paragraph>Common ocean birds are Brandt's Cormorant,
      Double-crested Cormorant, Pelagic Cormorant,
      Pigeon Guillemot, and the Tufted Puffin.
      </paragraph>
    </section>
    <section id="s1c">
      <title>Desert Birds</title>
      <paragraph>Birds you find in the desert include the
      Sage Grouse, California Quail, and Prairie Falcon.
      </paragraph>
    </section>
  </section>
</document>
```

Using this XML source file, we first extract the localizable XML text into XLIFF `<source>` elements within `<unit>` and `<segment>` elements (see Example 3.2).

Example 3.2 – XLIFF unit with one segment

```
<unit id="title-2">
  <segment>
    <source>Birds in Oregon</source>
  </segment>
</unit>
```

Optionally, we may also capture the XML document's format and metadata.

The container elements are the `<xliff>` root element and the `<file>` element. Example 3.3 shows the entire XLIFF file:

Example 3.3 – Initial XLIFF file for Example 3.1

```xml
<?xml version="1.0"?>
<xliff xmlns="urn:oasis:names:tc:xliff:document:2.0"
 xmlns:xmrk="urn:xmarker"
 xmlns:xsi="http://www.w3.org/2001/XMLSchema-instance"
 xsi:schemaLocation="xliff_core_2.0.xsd"
 version="2.0" srcLang="en" xml:lang="en">
<file id="myXML">
 <skeleton>
  <xmrk:document xmarker_idref="document-0" status="draft" id="d1">
   <xmrk:section xmarker_idref="section-1" id="s1">
    <xmrk:title xmarker_idref="title-2"/>
    <xmrk:paragraph xmarker_idref="paragraph-3"/>
    <xmrk:section xmarker_idref="section-4" id="s1a">
     <xmrk:title xmarker_idref="title-5"/>
     <xmrk:paragraph xmarker_idref="paragraph-6"/>
    </xmrk:section>
    <xmrk:section xmarker_idref="section-7" id="s1b">
     <xmrk:title xmarker_idref="title-8"/>
     <xmrk:paragraph xmarker_idref="paragraph-9"/>
    </xmrk:section>
    <xmrk:section xmarker_idref="section-10" id="s1c">
     <xmrk:title xmarker_idref="title-11"/>
     <xmrk:paragraph xmarker_idref="paragraph-12"/>
    </xmrk:section>
   </xmrk:section>
  </xmrk:document>
 </skeleton>
    <unit id="title-2">
      <segment>
        <source>Birds in Oregon</source>
      </segment>
    </unit>
    <unit id="paragraph-3">
      <segment>
       <source>Oregon is a mostly temperate state. There are many different
               kinds of birds that thrive there.</source>
      </segment>
    </unit>
    <unit id="title-5">
      <segment>
       <source>High Altitude Birds</source>
      </segment>
    </unit>
    <unit id="paragraph-6">
```

```
      <segment>
        <source>Birds that thrive in the high altitude include the
          White-tailed Ptarmigan, Sharp-tailed Grouse,
          Yellow-billed Loon, Cattle Egret, Gyrfalcon,
              Snowy Owl, Yellow-billed Cuckoo, and Boreal Owl.</source>
      </segment>
    </unit>
    <unit id="title-8">
      <segment>
        <source>Ocean Birds</source>
      </segment>
    </unit>
    <unit id="paragraph-9">
      <segment>
        <source>Common ocean birds are Brandt's Cormorant, Double-crested
              Cormorant, Pelagic Cormorant, Pigeon Guillemot, and the
          Tufted Puffin.</source>
      </segment>
    </unit>
    <unit id="title-11">
      <segment>
        <source>Desert Birds</source>
      </segment>
    </unit>
    <unit id="paragraph-12">
      <segment>
        <source>Birds you find in the desert include the Sage Grouse,
          California Quail, and Prairie Falcon.</source>
      </segment>
    </unit>
</file>
</xliff>
```

After transforming the source XML file to XLIFF, the file is ready to be translated. Example 3.4 shows the XLIFF file in Example 3.3 translated into Spanish.

Example 3.4 – Example 3.3 translated into Spanish

```
<?xml version="1.0"?>
<xliff xmlns="urn:oasis:names:tc:xliff:document:2.0"
       xmlns:xmrk="urn:xmarker"
       xmlns:xsi="http://www.w3.org/2001/XMLSchema-instance"
       xsi:schemaLocation="xliff_core_2.0.xsd" version="2.0" srcLang="en"
       trgLang="es" xml:lang="en">
  <file id="myXML">
    <skeleton>
      <xmrk:document xmarker_idref="document-0" status="draft"
      id="d1">
```

```
        <xmrk:section xmarker_idref="section-1" id="s1">
          <xmrk:title xmarker_idref="title-2" />
          <xmrk:paragraph xmarker_idref="paragraph-3" />
          <xmrk:section xmarker_idref="section-4" id="s1a">
            <xmrk:title xmarker_idref="title-5" />
            <xmrk:paragraph xmarker_idref="paragraph-6" />
          </xmrk:section>
          <xmrk:section xmarker_idref="section-7" id="s1b">
            <xmrk:title xmarker_idref="title-8" />
            <xmrk:paragraph xmarker_idref="paragraph-9" />
          </xmrk:section>
          <xmrk:section xmarker_idref="section-10" id="s1c">
            <xmrk:title xmarker_idref="title-11" />
            <xmrk:paragraph xmarker_idref="paragraph-12" />
          </xmrk:section>
        </xmrk:section>
      </xmrk:document>
    </skeleton>
    <unit id="title-2">
      <segment>
        <source>Birds in Oregon</source>
        <target>Pájaros en Oregon</target>
      </segment>
    </unit>
    <unit id="paragraph-3">
      <segment>
        <source>Oregon is a mostly temperate state. There are many
        different kinds of birds that thrive there.</source>
        <target>Oregon es un estado generalmente templado. Muchos
        tipos diferentes de pájaros prosperan allí.</target>
      </segment>
    </unit>
    <unit id="title-5">
      <segment>
        <source>High Altitude Birds</source>
        <target>Pájaros de gran altura</target>
      </segment>
    </unit>
    <unit id="paragraph-6">
      <segment>
        <source>Birds that thrive in the high altitude include the
        White-tailed Ptarmigan, Sharp-tailed Grouse, Yellow-billed
        Loon, Cattle Egret, Gyrfalcon, Snowy Owl, Yellow-billed
        Cuckoo, and Boreal Owl.</source>
        <target>Los pájaros que prosperan a grandes alturas
        incluyen a la perdiz nival de cola blanca, urogallo de las
        praderas, colimbo de Adams, garza boyera, halcón
        gerifalte, gran buho blanco, cuclillo piquigualdo y al
        mochuelo boreal.</target>
      </segment>
```

```
    </unit>
    <unit id="title-8">
      <segment>
        <source>Ocean Birds</source>
        <target>Pájaros oceánicos</target>
      </segment>
    </unit>
    <unit id="paragraph-9">
      <segment>
        <source>Common ocean birds are Brandt's Cormorant,
        Double-crested Cormorant, Pelagic Cormorant, Pigeon
        Guillemot, and the Tufted Puffin.</source>
        <target>Los pájaros oceánicos comunes son el cormorán de
        Brandt, cormorán orejudo, cormorán pelágico, arao
        colombino y el frailecillo coletudo.</target>
      </segment>
    </unit>
    <unit id="title-11">
      <segment>
        <source>Desert Birds</source>
        <target>Pájaros del desierto</target>
      </segment>
    </unit>
    <unit id="paragraph-12">
      <segment>
        <source>Birds you find in the desert include the Sage
        Grouse, California Quail, and Prairie Falcon.</source>
        <target>Los pájaros que se encuentran en el desierto
        incluyen al urogallo de las artemisas, codorniz
        californiana y al halcón pálido.</target>
      </segment>
    </unit>
  </file>
</xliff>
```

Example 3.5 shows the translated target file after the merger agent merges the XLIFF content and the XML format from the source file.

 There are many ways to transform a translated XLIFF file back to its original XML format. See Appendix B, *XSL Examples: Transforming Source to and from XLIFF* (p. 165), for examples of how to transform XML content to and from XLIFF using XSL.

Example 3.5 – Example 3.1 translated into Spanish

```
<document status="draft" id="d1">
  <section id="s1">
    <title>Pájaros en Oregon</title>
    <paragraph>Oregon es un estado generalmente templado. Muchos
    tipos diferentes de pájaros prosperan allí.</paragraph>
    <section id="s1a">
      <title>Pájaros de gran altura</title>
      <paragraph>Los pájaros que prosperan a grandes alturas
      incluyen a la perdiz nival de cola blanca, urogallo de las
      praderas, colimbo de Adams, garza boyera, halcón gerifalte,
      gran buho blanco, cuclillo piquigualdo y al mochuelo
      boreal.</paragraph>
    </section>
    <section id="s1b">
      <title>Pájaros oceánicos</title>
      <paragraph>Los pájaros oceánicos comunes son el cormorán
      de Brandt, cormorán orejudo, cormorán pelágico, arao
      colombino y el frailecillo coletudo.</paragraph>
    </section>
    <section id="s1c">
      <title>Pájaros del desierto</title>
      <paragraph>Los pájaros que se encuentran en el desierto
      incluyen al urogallo de las artemisas, codorniz californiana
      y al halcón pálido.</paragraph>
    </section>
  </section>
</document>
```

We use this generic workflow in the following sections for DITA content, web content, and office documents.

Tools for the transformation

Both commercial and open source tools are available that automate the transformation from any well-formed XML file to XLIFF and back. An example of an entry-level, open source tool is the xliffRoundTrip tool.[2]

---

[2] http://sourceforge.net/projects/xliffroundtrip/

CHAPTER 4
Translating DITA

The Darwin Information Typing Architecture (DITA) is an OASIS standard for producing structured XML content that you can use to generate multiple outputs from a single source. When properly used, DITA can help you reduce the amount of content that needs to be translated.

While single-sourcing is a strength in DITA, processing and managing topics and their relationships can add overhead, increase the risk of error, and increase costs. We will show how you can use XLIFF to manage these challenges.

A DITA project consists of set of XML files that can be classified into the following categories:

- **Topics:** Documents that can be published as parts of a larger set
- **Maps:** Documents that connect the topic files used in the project and organize the output
- **Value files:** Documents that contain variable values and filtering information

DITA includes two features that complicate the translation process: content references and specialization.

A content reference is an element that contains a `conref` attribute. This attribute points to content that will be pulled in to replace the element that contains the `conref` attribute. This mechanism enables the reuse of content in different parts of a project or in different projects.

DITA specialization provides a mechanism for defining new elements and attributes that extend the existing DITA schema. Specialization in DITA follows a set of rules that make it possible for DITA-aware processors to handle specialized content even if they don't have the specialized schema available.

DITA projects typically consist of many small files, which can complicate the translation process. DITA also provides the `translate` attribute, which identifies whether content should or should not be translated.

These characteristics of DITA mean that the standard methods used for translating XML documents are insufficient for translating DITA projects. The following sections describe what translation tools need to do to handle these characteristics.

Handling content references

The DITA file in Example 4.1 has conref attributes that reference elements from the file shown in Example 4.2. In Example 4.1, the elements that contain conref attributes are highlighted in bold text. In Example 4.2, the text pulled in using content references is highlighted.

Example 4.1 – DITA topic using the conref mechanism

```xml
<?xml version="1.0" encoding="UTF-8"?>
<!DOCTYPE task PUBLIC "-//OASIS//DTD DITA Task//EN" "task.dtd">
<task id="task_hdj_drv_bh">
  <title>Applying XSL Transformation</title>
  <taskbody>
    <steps>
      <step>
        <cmd>Open the document to transform.</cmd>
      </step>
      <step>
        <cmd>In the
          <uicontrol conref="ui_ref.dita#uiref/xsl_menu"/> menu,
          select
          <uicontrol conref="ui_ref.dita#uiref/xsl_trans"/>.
        </cmd>
      </step>
      <step>
        <cmd>Select the appropriate XSL Stylesheet</cmd>
      </step>
      <step>
        <cmd>Click the
          <uicontrol conref="ui_ref.dita#uiref/xsl_apply"/> button.
        </cmd>
      </step>
    </steps>
  </taskbody>
</task>
```

Example 4.2 – DITA topic file, `ui_ref.dita`, with text from Example 4.1

```
<?xml version="1.0" encoding="UTF-8"?>
<!DOCTYPE concept PUBLIC "-//OASIS//DTD DITA Concept//EN" "concept.dtd">
<concept id="uiref">
  <title>UI Elements</title>
  <conbody>
    <p><uicontrol id="xsl_menu">Transformation</uicontrol>:
      program menu that contains all transformation options.</p>
    <p><uicontrol id="xsl_trans">XSL Transformation</uicontrol>:
      applies an XSL Stylesheet to an XML document.</p>
    <p><uicontrol id="xsl_apply">Apply Transformation</uicontrol>:
      applies the selected XSL Stylesheet to the current open document.</p>
  </conbody>
</concept>
```

Technical writers working with DITA need an XML editor that resolves `conref` attributes and displays the referenced content in context.

Example 4.3 shows how a DITA-aware XML editor might display the text in Example 4.1

Example 4.3 – Example 4.1 displayed in a DITA-aware editor

1. Open the document to transform.
2. In the *Transformation* menu, select *XSL Transformation*.
3. Click the *Apply Transformation* button.

The translator must be able to see the text being translated in a complete representation. If not, he or she will not have the necessary context to translate that text.

Example 4.4 shows an XLIFF 2.0 file that contains the text extracted from Example 4.1 with all references from Example 4.2 resolved. As with previous examples, the text pulled in using content references is highlighted in bold text.

Example 4.4 – XLIFF file build from Example 4.1 and Example 4.2

```
<unit id="1">
  <segment>
    <source>Applying XSL Transformation</source>
  </segment>
</unit>
<unit id="2">
  <segment>
    <source>Open the document to transform.</source>
  </segment>
</unit>
<unit id="3">
  <segment>
    <source>In the <pc id="1">Transformation</pc> menu,
      select  <pc id="2">XSL Transformation</pc>.</source>
  </segment>
</unit>
<unit id="6">
  <segment>
    <source>Select the appropriate XSL Stylesheet</source>
  </segment>
</unit>
<unit id="4">
  <segment>
    <source>Click the <pc id="1">Apply Transformation</pc> button.
    </source>
  </segment>
</unit>
```

Handling DITA specialization

DITA includes a set of DTDs and XML Schemas that contain most elements and attributes needed in a standard documentation project. Nevertheless, sometimes you may find that the standard set of elements and attributes is not enough and that you need custom extensions.

The DITA specialization mechanism lets you modify the default DITA schemas to incorporate additional elements and attributes.

Unfortunately, not all tools support DITA specialization. If you use DITA specialization, you need a translation tool that will allow you to do the following:

- Customize the list of translatable elements and attributes
- Incorporate custom DTDs and XML Schemas

Even if you don't use specialization, you may still need customized translations. For example, the `<draft-comment>` element is normally used internally, and readers of the published document-ation almost never see its content. Therefore, the `<draft-comment>` element is usually treated as untranslatable by CAT tools. However, you may need to translate comments for reviewers. You will need a customizable CAT tool to do this.

Using the translate attribute

DITA provides the translate attribute, which lets you mark portions of your text as translatable or untranslatable. To mark the contents of an element as untranslatable, set the value of the translate attribute to no, as shown in Example 4.5.

Example 4.5 – DITA `translate` attribute on text that should not be translated

```
<p translate="no">Warning: this text should not be translated.</p>
```

Some translation tools ignore the `translate` attribute and extract the text anyway.

The `translate` attribute should be used with block level elements (those that contain full para-graphs or sentences) such as <p>. Setting the `translate` attribute to "no" in inline elements (those that appear in the middle of a sentence) is not a good idea because a translator working with the surrounding text needs to see the element content for context. Example 4.6 shows how you can safely protect untranslatable text that appears in the middle of a sentence by referencing a copy stored in an untranslatable element.

Example 4.6 – Untranslatable inline text protected in `<draft-comment>`

```
<?xml version="1.0" encoding="UTF-8"?>
<!DOCTYPE concept PUBLIC "-//OASIS//DTD DITA Concept//EN" "concept.dtd">
<concept id="locking">
  <title translate="no">Untranslatable Title</title>
  <conbody>
    <p>This sentence contains <ph conref="#locking/lock"/> text.</p>
    <draft-comment translate="no">
      <ph id="lock">untranslatable</ph>
    </draft-comment>
  </conbody>
</concept>
```

A translation tool parsing Example 4.6 should be able to do the following:

- Ignore the `<title>` element
- Include the word "untranslatable" when extracting the `<p>` element
- Ignore the `<draft-comment>` element

Managing many small files

A DITA project may contain hundreds of small files. That's not unusual, but it does make file handling more difficult.

When working with a large number of files, many DITA teams use a Content Management System (CMS). Although a CMS is not required for working with DITA, having one can simplify project management.

A CMS can help you separate the files referenced by a DITA map and prepare a package for translation. Your CMS may also be able to identify which files require translation and create a translation package that includes only those files.

If you don't have a CMS, a DITA-enabled translation tool should be able to parse a DITA map, resolve the references, identify which files need translation, and prepare a unified package for your LSP.

If your LSP charges you for file management, you can reduce cost by preparing a consolidated translation package in house.

CHAPTER 5
Translating Websites

Whether simple or complex, websites can be challenging to translate because they usually contain many different resources.

In this section, we first provide an example of a simple website that includes only HTML files and linked PNG graphic files. Then, we provide an example of a complex web site that is managed by a Web Content Management System (Web CMS), Drupal, and delivers content from a database.

Translating simple HTML websites

Simple websites, made of HTML files and graphics, can be challenging to translate. The translation workflow requires you to process multiple HTML files in the following ordere:

1. Gather the content to be translated
2. Package the content
3. Supply the packaged content to the translator
4. Leverage content against translation memory
5. Translate the content
6. Review the translation in-context
7. Approve the translation
8. Publish the translated website

Even in simple HTML websites, where most translatable text comes from HTML elements, translatable text can sometimes come from other sources in the HTML. For example, sometimes attributes contain text that needs to be translated, such as the `alt` attribute on images or the `title` attribute, which is commonly used for tool tips. In some cases, translatable text can occur in comments or CDATA sections. Text from all these sources needs to be extracted into `<unit>` elements. However, for this simple HTML website workflow example, we assume the translatable text comes only from elements.

Using XLIFF to manage a simple HTML-website translation workflow solves many complexities. Each of the HTML files can be captured in an separate `<file>` element. In XLIFF, the source structure or format for each HTML file is captured in a skeleton, and the translatable text is captured in `<unit>` and `<segment>` elements.

Consider the website diagrammed in Figure 5.1. The HTML for the files on this site is shown in Example 5.1 through Example 5.6.

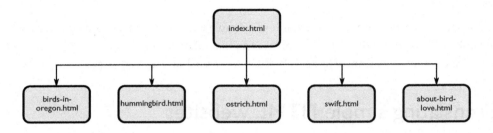

Figure 5.1 – Simple HTML website

Example 5.1 – HTML website: `index.html`

```html
<html>
  <head></head>
  <body>
    <h1>Bird Love</h1>
    <table>
      <tr>
        <td>
          <a href="birds-in-oregon.html">Birds in Oregon</a>
        </td>
        <td>
          <a href="hummingbird.html">Hummingbird</a>
        </td>
        <td>
          <a href="ostrich.html">Ostrich</a>
        </td>
        <td>
          <a href="swift.html">Swift</a>
        </td>
        <td>
          <a href="about-bird-love.html">About Bird Love</a>
        </td>
      </tr>
    </table>
  </body>
</html>
```

Example 5.2 – HTML website: `birds-in-oregon.html`

```html
<html id="d1">
  <head></head>
  <body>
    <div id="s1">
      <h2>Birds in Oregon</h2>
      <p>Oregon is a mostly temperate state. There are many
      different kinds of birds that thrive there.</p>
      <div id="s1a">
        <h2>High Altitude Birds</h2>
        <p>Birds that thrive in high altitude include the
        White-tailed Ptarmigan, Sharp-tailed Grouse, Yellow-billed
        Loon, Cattle Egret, Gyrfalcon, Snowy Owl, Yellow-billed
        Cuckoo, and Boreal Owl.</p>
      </div>
      <div id="s1b">
        <h2>Ocean Birds</h2>
        <p>Common ocean birds are Brandt's Cormorant,
        Double-crested Cormorant, Pelagic Cormorant, Pigeon
        Guillemot, and Tufted Puffin.</p>
      </div>
      <div id="s1c">
        <h2>Desert Birds</h2>
        <p>Birds you find in the desert include the Sage Grouse,
        California Quail, and Prairie Falcon.</p>
      </div>
    </div>
  </body>
</html>
```

Example 5.3 – HTML website: `hummingbird.html`

```html
<html id="hummingbird"
      xml:lang="en-US">
  <head></head>
  <body>
    <h2>Hummingbird</h2>
    <p>Smallest bird: Bee hummingbird (2-1/4 in)</p>
  </body>
</html>
```

Example 5.4 – HTML website: `ostrich.html`

```
<html id="ostrich"
      xml:lang="en-US">
  <head></head>
  <body>
    <h2>Ostrich</h2>
    <p>Heaviest bird: Ostrich (330 lb)</p>
  </body>
</html>
```

Example 5.5 – HTML website: `swift.html`

```
<html id="swift"
      xml:lang="en-US">
  <head></head>
  <body>
    <h2>Swift</h2>
    <p>Fastest bird flying: Common Swift (125 mi/hr)</p>
  </body>
</html>
```

Example 5.6 – HTML website: `about-bird-love.html`

```
<html id="abl"
      xml:lang="en-US">
  <head></head>
  <body>
    <h2>About Bird Love</h2>
    <p>Bird Love is an organization that exists for the love of
    birds.</p>
  </body>
</html>
```

The website is extracted into the XLIFF file shown in Example 5.7. The text from each HTML file is contained in a separate `<file>` element in the XLIFF file.

There are many ways to transform this simple HTML to XLIFF. See the section titled "Transforming to and from web XLIFF using XSLT" (p. 165) for one XSLT solution for transforming HTML to XLIFF.

Example 5.7 – XLIFF file for the simple HTML website

```
<xliff>
  <file id="f1">
    <unit name="xmrk:h1d1e2" id="d01">
      <segment>
        <source>Bird Love</source>
      </segment>
    </unit>
    <unit name="xmrk:ad1e7" id="d14">
      <segment>
        <source>Birds in Oregon</source>
      </segment>
    </unit>
    <unit name="xmrk:ad1e10" id="d34">
      <segment>
        <source>Hummingbird</source>
      </segment>
    </unit>
    <unit name="xmrk:ad1e13" id="d54">
      <segment>
        <source>Ostrich</source>
      </segment>
    </unit>
    <unit name="xmrk:ad1e16" id="d74">
      <segment>
        <source>Swift</source>
      </segment>
    </unit>
    <unit name="xmrk:ad1e19" id="d94">
      <segment>
        <source>About Bird Love</source>
      </segment>
    </unit>
  </file>
  <file id="f2">
    <unit name="xmrk:h2d2e3" id="d02">
      <segment>
        <source>Birds in Oregon</source>
      </segment>
    </unit>
    <unit name="xmrk:pd2e5" id="d12">
      <segment>
        <source>Oregon is a mostly temperate state. There are many
        different kinds of birds that thrive there.</source>
      </segment>
    </unit>
    <unit name="xmrk:h2d2e8" id="d23">
      <segment>
        <source>High Altitude Birds</source>
```

```
        </unit>
        <unit name="xmrk:pd2e10" id="d33">
          <segment>
            <source>Birds that thrive in the high altitude include the
            White-tailed Ptarmigan, Sharp-tailed Grouse, Yellow-billed
            Loon, Cattle Egret, Gyrfalcon, Snowy Owl, Yellow-billed
            Cuckoo, and Boreal Owl.</source>
          </segment>
        </unit>
        <unit name="xmrk:h2d2e13" id="d53">
          <segment>
            <source>Ocean Birds</source>
          </segment>
        </unit>
        <unit name="xmrk:pd2e15" id="d63">
          <segment>
            <source>Common ocean birds are Brandt's Cormorant,
            Double-crested Cormorant, Pelagic Cormorant, Pigeon
            Guillemot, and Tufted Puffin.</source>
          </segment>
        </unit>
        <unit name="xmrk:h2d2e18" id="d83">
          <segment>
            <source>Desert Birds</source>
          </segment>
        </unit>
        <unit name="xmrk:pd2e20" id="d93">
          <segment>
            <source>Birds you find in the desert include the Sage
            Grouse, California Quail, and Prairie Falcon.</source>
          </segment>
        </unit>
      </file>
      <file id="f3">
        <unit name="xmrk:h2d3e2" id="d01">
          <segment>
            <source>Hummingbird</source>
          </segment>
        </unit>
        <unit name="xmrk:pd3e4" id="d11">
          <segment>
            <source>Smallest bird: Bee hummingbird (2-1/4 in)</source>
          </segment>
        </unit>
      </file>
      <file id="f4">
        <unit name="xmrk:h2d4e2" id="d01">
          <segment>
            <source>Ostrich</source>
```

```
      </unit>
      <unit name="xmrk:pd4e4" id="d11">
        <segment>
          <source>Heaviest bird: Ostrich (330 lb)</source>
        </segment>
      </unit>
    </file>
    <file id="f5">
      <unit name="xmrk:h2d5e2" id="d01">
        <segment>
          <source>Swift</source>
        </segment>
      </unit>
      <unit name="xmrk:pd5e4" id="d11">
        <segment>
          <source>Fastest bird flying: Common Swift (125
          mi/hr)</source>
        </segment>
      </unit>
    </file>
    <file id="f6">
      <unit name="xmrk:h2d6e2" id="d01">
        <segment>
          <source>About Bird Love</source>
        </segment>
      </unit>
      <unit name="xmrk:pd6e4" id="d11">
        <segment>
          <source>Bird Love is an organization that exists for the
          love of birds.</source>
        </segment>
      </unit>
    </file>
</xliff>
```

Example 5.8 shows the XLIFF file after it has been translated to Spanish.

Example 5.8 – XLIFF file from Example 5.7 translated into Spanish

```
<xliff>
  <file id="f1">
    <unit name="xmrk:h1d1e2" id="d01">
      <segment>
        <source>Bird Love</source>
        <target>Amor por los pájaros</target>
      </segment>
    </unit>
```

```xml
<unit name="xmrk:ad1e7" id="d14">
  <segment>
    <source>Birds in Oregon</source>
    <target>Pájaros en Oregon</target>
  </segment>
</unit>
<unit name="xmrk:ad1e10" id="d34">
  <segment>
    <source>Hummingbird</source>
    <target>Colibrí</target>
  </segment>
</unit>
<unit name="xmrk:ad1e13" id="d54">
  <segment>
    <source>Ostrich</source>
    <target>Avestruz</target>
  </segment>
</unit>
<unit name="xmrk:ad1e16" id="d74">
  <segment>
    <source>Swift</source>
    <target>Vencejo</target>
  </segment>
</unit>
<unit name="xmrk:ad1e19" id="d94">
  <segment>
    <source>About Bird Love</source>
    <target>Acerca de "Amor por los pájaros"</target>
  </segment>
</unit>
</file>
<file id="f2">
  <unit name="xmrk:h2d2e3" id="d02">
    <segment>
      <source>Birds in Oregon</source>
      <target>Pájaros en Oregon</target>
    </segment>
  </unit>
  <unit name="xmrk:pd2e5" id="d12">
    <segment>
      <source>Oregon is a mostly temperate state. There are many
      different kinds of birds that thrive there.</source>
      <target>Oregon es un estado generalmente templado. Muchos
      tipos diferentes de pájaros prosperan allí.</target>
    </segment>
  </unit>
  <unit name="xmrk:h2d2e8" id="d23">
    <segment>
      <source>High Altitude Birds</source>
      <target>Pájaros de gran altura</target>
```

```
    </unit>
    <unit name="xmrk:pd2e10" id="d33">
      <segment>
        <source>Birds that thrive in the high altitude include the
        White-tailed Ptarmigan, Sharp-tailed Grouse, Yellow-billed
        Loon, Cattle Egret, Gyrfalcon, Snowy Owl, Yellow-billed
        Cuckoo, and Boreal Owl.</source>
        <target>Los pájaros que prosperan a grandes alturas incluyen
        a la perdiz nival de cola blanca, urogallo de las praderas,
        colimbo de Adams, garza boyera, halcón gerifalte, gran buho
        blanco, cuclillo piquigualdo y al mochuelo boreal.</target>
      </segment>
    </unit>
    <unit name="xmrk:h2d2e13" id="d53">
      <segment>
        <source>Ocean Birds</source>
        <target>Pájaros oceánicos</target>
      </segment>
    </unit>
    <unit name="xmrk:pd2e15" id="d63">
      <segment>
        <source>Common ocean birds are Brandt's Cormorant,
        Double-crested Cormorant, Pelagic Cormorant, Pigeon
        Guillemot, and Tufted Puffin.</source>
        <target>Los pájaros oceánicos comunes son el cormorán de
        Brandt, cormorán orejudo, cormorán pelágico, arao
        colombino y el frailecillo coletudo.</target>
      </segment>
    </unit>
    <unit name="xmrk:h2d2e18" id="d83">
      <segment>
        <source>Desert Birds</source>
        <target>Pájaros del desierto</target>
      </segment>
    </unit>
    <unit name="xmrk:pd2e20" id="d93">
      <segment>
        <source>Birds you find in the desert include the Sage Grouse,
        California Quail, and Prairie Falcon.</source>
        <target>Los pájaros que se encuentran en el desierto
        incluyen al urogallo de las artemisas, codorniz californiana
        y al halcón pálido.</target>
      </segment>
    </unit>
  </file>
  <file id="f3">
    <unit name="xmrk:h2d3e2" id="d01">
      <segment>
        <source>Hummingbird</source>
```

```
      <target>Colibrí</target>
  </unit>
  <unit name="xmrk:pd3e4" id="d11">
    <segment>
      <source>Smallest bird: Bee hummingbird (2-1/4 in)</source>
      <target>Pájaro más pequeño: Colibrí zunzuncito (5
      cm)</target>
    </segment>
  </unit>
</file>
<file id="f4">
  <unit name="xmrk:h2d4e2" id="d01">
    <segment>
      <source>Ostrich</source>
      <target>Avestruz</target>
    </segment>
  </unit>
  <unit name="xmrk:pd4e4" id="d11">
    <segment>
      <source>Heaviest bird: Ostrich (330 lb)</source>
      <target>Pájaro más pesado: Avestruz (150 Kg)</target>
    </segment>
  </unit>
</file>
<file id="f5">
  <unit name="xmrk:h2d5e2" id="d01">
    <segment>
      <source>Swift</source>
      <target>Vencejo</target>
    </segment>
  </unit>
  <unit name="xmrk:pd5e4" id="d11">
    <segment>
      <source>Fastest bird flying: Common Swift (125
      mi/hr)</source>
      <target>Pájaro de vuelo más veloz: Vencejo común (200
      Km/h)</target>
    </segment>
  </unit>
</file>
<file id="f6">
  <unit name="xmrk:h2d6e2" id="d01">
    <segment>
      <source>About Bird Love</source>
      <target>Acerca de "Amor por los pájaros"</target>
    </segment>
  </unit>
  <unit name="xmrk:pd6e4" id="d11">
    <segment>
```

```
      <source>Bird Love is an organization that exists for the love
      of birds.</source>
      <target>"Amor por los pájaros" es una organización que
      existe por amor a los pájaros.</target>
  </unit>
  </file>
</xliff>
```

After transforming the translated XLIFF back to HTML, we get the translated HTML website files shown in Example 5.9 through Example 5.14.

Many methods are available to transform the translated XLIFF back to HTML. See the section titled "Transforming to and from web XLIFF using XSLT" (p. 165) for an XSLT transformation that will do the job.

Example 5.9 – HTML website in Spanish: `index.html`

```
<html>
  <head></head>
  <body>
    <h1>Amor por los pájaros</h1>
    <table>
      <tr>
        <td>
          <a href="birds-in-oregon.html">Pájaros en Oregon</a>
        </td>
        <td>
          <a href="hummingbird.html">Colibrí</a>
        </td>
        <td>
          <a href="ostrich.html">Avestruz</a>
        </td>
        <td>
          <a href="swift.html">Vencejo</a>
        </td>
        <td>
          <a href="about-bird-love.html">Acerca de "Amor por los
          pájaros"</a>
        </td>
      </tr>
    </table>
  </body>
</html>
```

Example 5.10 – HTML website in Spanish: `birds-in-oregon.html`

```
<html id="d1">
  <head></head>
  <body>
    <div id="s1">
      <h2>Pájaros en Oregon</h2>
      <p>Oregon es un estado generalmente templado. Muchos tipos
      diferentes de pájaros prosperan allí.</p>
      <div id="s1a">
        <h2>Pájaros de gran altura</h2>
        <p>Los pájaros que prosperan a grandes alturas incluyen a
        la perdiz nival de cola blanca, urogallo de las praderas,
        colimbo de Adams, garza boyera, halcón gerifalte, gran
        buho blanco, cuclillo piquigualdo y al mochuelo boreal.</p>
      </div>
      <div id="s1b">
        <h2>Pájaros oceánicos</h2>
        <p>Los pájaros oceánicos comunes son el cormorán de
        Brandt, cormorán orejudo, cormorán pelágico, arao
        colombino y el frailecillo coletudo.</p>
      </div>
      <div id="s1c">
        <h2>Pájaros del desierto</h2>
        <p>Los pájaros que se encuentran en el desierto incluyen
        al urogallo de las artemisas, codorniz californiana y al
        halcón pálido.</p>
      </div>
    </div>
  </body>
</html>
```

Example 5.11 – HTML website in Spanish: `hummingbird.html`

```
<html id="hummingbird"
      xml:lang="en-US">
  <head></head>
  <body>
    <h2>Colibrí</h2>
    <p>Pájaro más pequeño: Colibrí zunzuncito (5 cm)</p>
  </body>
</html>
```

Example 5.12 – HTML website in Spanish: `ostrich.html`

```
<html id="ostrich"
      xml:lang="en-US">
  <head></head>
  <body>
    <h2>Avestruz</h2>
    <p>Pájaro más pesado: Avestruz (150 Kg)</p>
  </body>
</html>
```

Example 5.13 – HTML website in Spanish: `swift.html`

```
<html id="swift"
      xml:lang="en-US">
  <head></head>
  <body>
    <h2>Vencejo</h2>
    <p>Pájaro de vuelo más veloz: Vencejo común (200 Km/h)</p>
  </body>
</html>
```

Example 5.14 – HTML website in Spanish: `about-bird-love.html`

```
<html id="abl"
      xml:lang="en-US">
  <head></head>
  <body>
    <h2>Acerca de "Amor por los pájaros"</h2>
    <p>"Amor por los pájaros" es una organización que existe por
    amor a los pájaros.</p>
  </body>
</html>
```

Translating complex websites

Websites can be complex for several reasons. One reason is sheer volume. Large websites are often managed by a Web CMS. In such cases, there may be no physical files to translate. Instead, a database stores content and renders web pages as needed. In addition, translatable text may be found in bitmap images, flash files, and asp files, to name just a few.

In this section, we will look at a typical database-driven website managed by Drupal. Drupal is an open source CMS that uses Apache, MySQL, and PHP to store content in a database and render it for web visitors.

Most Web CMS tools have a translation API, and some have built-in XLIFF support. Drupal includes a robust XLIFF module, which extracts nodes from the database and transforms them into XLIFF. After the XLIFF is translated, the XLIFF module imports the translated XLIFF back into the database, where it can be used by Drupal to create a localized site.

The Drupal XLIFF module is good, but not perfect. For example, the module may not retrieve all translatable strings. Some strings must be handled as Portable Objects (PO) files.[1] In our example, we assume no translations are needed for external files like flash or bitmap images.

Translating Web CMS: Drupal

We will use the Drupal XLIFF module to extract content from the database and create an XLIFF file for the translator to use. We will then use the same module to import the translated content back into the database and identified it to Drupal so it can be rendered correctly in the target language.

 Drupal XLIFF module

> You can download the Drupal XLIFF Tools from http://drupal.org/project/xliff. We used this downloadable version[2] in the following examples.

1. Log into Drupal.
2. Select the **Content management** menu, which appears at the top of the screen, above the main navigation bar.

Content management   Site building   Site configuration   Messaging   User management   Reports   Rules

Figure 5.2 – Drupal CMS menu

3. On the **Content management** menu, select **XLIFF tools → Settings**.

---

[1] http://www.gnu.org/software/gettext/manual/html_node/PO-Files.html

[2] http://drupal.org/files/xliff.zip

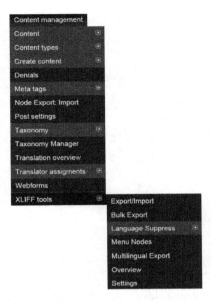

Figure 5.3 – Drupal **Content management** menu

4. On **Settings** menu, select the node type you want to translate.

Each node type refers to a category of information on the website. In this example, we selected **Species**. That means that any Drupal node that conforms to this type will be included in the XLIFF export.

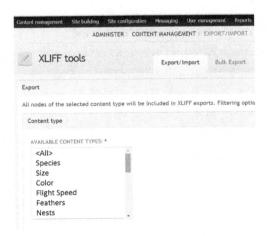

5. On the **Content management** menu, select **XLIFF tools** → **Export/Import**.

6. Choose **Export All** and click **Export as XLIFF**.

7. Save the `Species.xlf` XLIFF file locally.

8. If necessary, apply post processing to ensure that database fields are not mistakenly translated. For example, translating value of the **inventory status** field would corrupt the Drupal database. We post process the content, adding the annotation *do not translate* to protect that field.

Example 5.15 – `Species.xlf`

```
<unit id="field_warranty-13">
    <source>One-year Warranty</source>
    <target>One-year Warranty</target>
</unit>
<unit id="field_inventory_status-15">
    <source>Active</source>
    <target><mrk id="mrk1" translate="no">Active</mrk></target>
</unit>
```

9. Translate the `Species.xlf` XLIFF file (see Example 5.15).

Example 5.16 – `Species.xlf` translated

```
<unit id="field_warranty-13">
    <source>One-year Warranty</source>
    <target> 1-Jahres-Garantie </target>
</unit>
<unit id="field_inventory_status-15">
    <source>Active</source>
    <target><mrk id="mrk1" translate="no">Active</mrk></target>
</unit>
```

10. On the **Content management** menu, select **XLIFF tools** → **Export/Import**. Browse to the translated `Species.xlf` XLIFF file and click **Import XLIFF**.

11. In some cases, non-content text strings require additional processing. For example, in Drupal, menu strings are not typically processed in the Drupal XLIFF module.

Figure 5.4 – Drupal text not processed by the Drupal XLIFF module

On the **Site building** menu, select **Translate interface** → **Export**.

12. On the **Translate interface → Export** screen, select **Menu Export**. This will select the non-content text strings in this example.

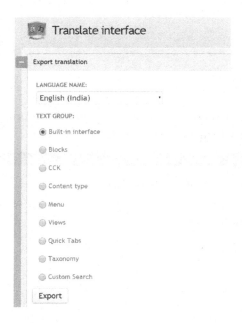

13. Save the menu.po file locally. This file contains Drupal UI interface strings and provides a non-XLIFF method to translate the menu text (see Example 5.17).

Example 5.17 – menu.po

```
#: item:1195:title
msgid "Home"
msgstr ""

#: item:19232:title
msgid "About Us"
msgstr ""

#: item:19250:title
msgid "Service"
msgstr ""
```

14. Convert the menu.po files to an XLIFF file (see Example 5.18).

Example 5.18 – XLIFF file for fragment for menu.po

```
<unit id="d_16">
 <segment>
  <source>Home</source>
  <target>Home</target>
 </segment>
</unit>
<unit id="d_20">
 <segment>
  <source>About Us</source>
  <target>About Us</target>
 </segment>
</unit>
<unit id="d_24">
 <segment>
  <source>Service</source>
  <target>Service</target>
 </segment>
</unit>
```

15. Translate the XLIFF files.

16. Merge translated XLIFF to menu.po using your translation software. Example 5.19 shows the translated XLIFF file.

Example 5.19 – XLIFF file translated into German

```
<unit id="d_16">
 <segment>
  <source>Home</source>
  <target> Startseite</target>
 </segment>
</unit>
<unit id="d_20">
 <segment>
  <source>About Us</source>
  <target> Über uns</target>
 </segment>
</unit>
<unit id="d_24">
 <segment>
  <source>Service</source>
  <target>Service</target>
 </segment>
</unit>
```

Example 5.20 shows the translated version of menu.po.

Example 5.20 – menu.po translated into German

```
#: item:1195:title
msgid "Home"
msgstr " Startseite"

#: item:19232:title
msgid "About Us"
msgstr " Über uns"

#: item:19250:title
msgid "Service"
msgstr "Service"
```

17. On the **Translate interface** → **Import** screen, select **Menu Import**.

With translated XLIFF content and XLIFF PO imported, the website is translated.

Figure 5.5 – Translated website

CHAPTER 6

Translating Office Documents

Microsoft Office and OpenOffice.org are two popular office products. These products work with three different document formats:

- XML-based formats
- Binary formats
- Rich Text Format (RTF)

XML-based formats

Since the release of Microsoft Office 2007, the default format for spreadsheets, charts, presentations, and word processing documents is Office Open XML (informally known as OOXML or OpenXML). The Office Open XML format is an open standard, originally published by Ecma as ECMA-376 and later by ISO and IEC as ISO/IEC 29500.

OpenOffice.org uses the OpenDocument format for all of its document types. OpenDocument was initially developed by Sun Microsystems and then made public as part of the OpenOffice.org project. OpenDocument is an OASIS standard, and version 1.1 is also an ISO/IEC international standard, ISO/IEC 26300:2006.

Both formats, Office Open XL and OpenDocument, store all parts of a document in zip files. Text and layout information are stored in XML files inside the zip package, along with any graphics or multimedia.

The basic recipe for translating XML-based office documents is as follows:

1. Decompress the zip package
2. Convert all contained XML files to XLIFF
3. Merge all XLIFF files into a unified XLIFF document
4. Translate the unified XLIFF
5. Split the translated, unified XLIFF, recreating the original set of XLIFF files
6. Convert the set of XLIFF files back to the original XML format
7. Zip the translated XML files, reconstructing the package

Although Office Open XML is based on zipped XML, it has some differences in the way formatting information is represented that require a special filter for step 2. For example, the most common way to represent inline formatting in an XML document is to enclose portions of text in elements that represent layout, such as shown in Example 6.1.

Example 6.1 – HTML inline markup

```
<p>This is <strong>strong</strong> text.
```

Microsoft uses a different approach. Instead of enclosing text in an element, it places marker elements at the beginning and end of the formatted section. Example 6.2 does not use actual Office Open XML markup, but it gives you an idea of how this approach works.

Example 6.2 – Office Open XML inline markup

```
<para>This sentence has <startBold/>bold<endBold/> text.</para>
```

Microsoft's approach has the advantage that it allows overlapping formatting, such as in Example 6.3.

Example 6.3 – Overlapping inline markup

```
<para>This sentence has <startBold/>bold,
  <startItalics/> bold and italics <endBold/>
  and just italics<endItalics/> text.
</para>
```

However, it has the disadvantage that text marked this way requires specialized filters that can handle the following cases:

- Documents with hidden text
- Documents with revision marks and comments
- Documents with other documents embedded (for example, an Excel sheet embedded in a Word document)

Binary office format

Before the release of Microsoft Office 2007, Microsoft Office stored documents using a proprietary binary format. Only a few programs could read those binary files because the specifications were closed, and only a handful of Microsoft partners knew how to handle them. Microsoft published the documentation in 2008.

You can find commercial libraries that will extract translatable text from binary Office files. However, it is more convenient to convert those binary files to the more friendly XML-based Office Open XML format and extract translatable text from the converted versions.

Rich Text Format

Rich Text Format (commonly known as RTF) is a proprietary document format for formatted text developed and maintained by Microsoft. RTF specifications are published by Microsoft and are available for free use.

RTF serves as a common format for exchanging formatted documents between word processing software (most word processors support export/import of documents in this format) and for migrating content from one operating system to another.

RTF documents are written in plain text, using either the ASCII or UTF-8 character set. Formatting is defined using special control commands preceded by the backslash ("\") character. Braces ("{" and "}") are used to enclose document sections or parts with similar formatting.

Example 6.4 – RTF fragment

```
\pard\pardeftab720\ri0

\f0\fs24 \cf0 Sample sentence with
\b bold
\b0  and
\i italicized\i0  text.
\f1 \
}}
```

An application would render the RTF fragment in Example 6.4 as shown here:

Sample sentence with **bold** and *italicized* text.

There are differences between RTF files produced by Word for Windows and Word for Mac OS X, especially regarding fonts and support for non-ASCII characters. OpenOffice.org uses its own encoding scheme for storing non-ASCII characters, which is also supported by Microsoft's tools.

Extracting translatable text from RTF documents requires specialized filters. There are commercial and open source tools capable of generating XLIFF files from RTF documents.

Unclean RTF

Some translation tools designed as macros for Microsoft Word work with bilingual RTF documents, commonly called *unclean* files. Unclean RTF was one of the first bilingual formats used for exchanging content in the translation industry.

An unclean RTF file looks like this in Microsoft Word:

{0>Sample sentence with **bold** and *italicized* text.<}100{>Oración de ejemplo con texto en **negrita** y *cursiva*.<0}

Unclean files cannot always be exchanged between different translation tools, especially between different operating systems. Exchange problems also occur when users have different versions of Word or when their versions have user interfaces in different languages. Converting unclean files to XLIFF solves these exchange problems.

An XLIFF filter that deals with unclean RTF files should handle the following cases:

- Partially segmented text
- Partially translated text
- Text with fuzzy matches from a translation memory engine

CHAPTER 7
Translating Graphics

Translating text in graphics has a long history of causing dread in technical documentation and translation practitioners. Technical publishers and LSPs can translate text in graphics, but it is expensive and error-prone. To illustrate the difficulty, consider two traditional methods for translating text in graphics.

- **Traditional method 1:** Manually extract text strings from the graphics and paste them into a spreadsheet.
- **Traditional method 2:** Send the graphics to the LSP and expect them to develop in-house expertise to process the graphic format.

Traditional method 1

Transfer strings in a spreadsheet, using the following steps:

1. Manually extract strings from each graphic
2. Provide the text strings to the translator in a word processor or spreadsheet file
3. The translator translates the text strings
4. Paste the translated strings back into the graphics file

Figure 7.1 illustrates this method. As you can see, this method adds a lot of overhead for the project manager and requires extensive involvement from the translator, graphic artist, and technical writer, who often end up taking shortcuts.

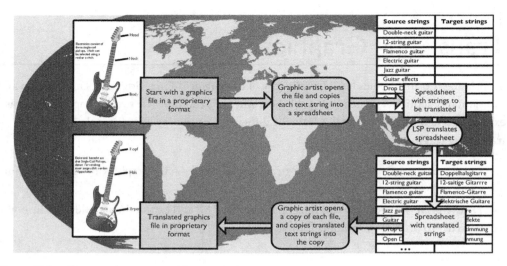

Figure 7.1 – Copy and paste graphic translation process

For example, as shown in Figure 7.2, a writer unfamiliar with the target language may receive the translated spreadsheet and accidentally drop a character when pasting the text into the graphic. And, because the writer doesn't know the target language, the mistake may go unnoticed.

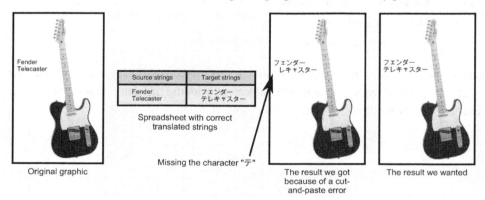

Figure 7.2 – Copy and paste example with a cut-and-paste error

Traditional method 2

The writer or illustrator gathers all of the images in their source format and sends them to the translator. The translator opens each graphic with a tool that will edit the graphic. The translator translates the text and returns the files to the writer.

The main drawback of this method is that the translator must have expertise in the graphics tool in addition to being a translator. This takes skills beyond translation. It will be more expensive and still error-prone.

Using SVG for translating graphics with XLIFF

SVG (Scalable Vector Graphics)[1] is an XML language for describing two-dimensional graphics and graphics applications. SVG graphics are easily rendered, created, and edited using any of several widely available software packages. SVG files are XML, so the text strings can be easily identified, extracted, and transformed.

The SVG process relies on the extract and merge method described in the section titled "The translation process: extract and merge" (p. 4). Here are the steps:

1. Begin with an SVG file (Figure 7.3).

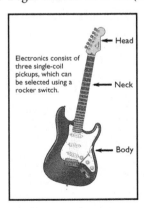

Figure 7.3 – Rendered SVG File

2. The translatable strings are accessible as SVG <tspan> elements.

---

[1] http://www.w3.org/Graphics/SVG/

Example 7.1 – Text elements for Figure 7.3

```
<text transform="matrix(1 0 0 1 249 29.3076)">
  <tspan x="0" y="0" font-family="'Gill Sans'" font-size="12">Head</tspan>
</text>
  <text transform="matrix(1 0 0 1 243 116.3076)">
  <tspan x="0" y="0" font-family="'Gill Sans'" font-size="12">Neck</tspan>
</text>
  <text transform="matrix(1 0 0 1 243 227.8076)">
  <tspan x="0" y="0" font-family="'Gill Sans'" font-size="12">Body</tspan>
</text>
<text transform="matrix(1 0 0 1 0 92.3076)">
  <tspan x="0" y="0" font-family="'Gill Sans'"
         font-size="12">Electronics consist of </tspan>
  <tspan x="0" y="14.4" font-family="'Gill Sans'"
         font-size="12">three single-coil </tspan>
  <tspan x="0" y="28.8" font-family="'Gill Sans'"
         font-size="12">pickups, which can </tspan>
  <tspan x="0" y="43.2" font-family="'Gill Sans'"
         font-size="12">be selected using a </tspan>
  <tspan x="0" y="57.6" font-family="'Gill Sans'"
         font-size="12">rocker switch.</tspan>
</text>
```

3. Extract the text strings into XLIFF <unit> elements.

Example 7.2 – XLIFF file for Example 7.1

```
<unit id="d0e445" name="x-tspan" xmrk:locate_id="0">
   <segment>
     <source>Head</source>
   </segment>
</unit>
<unit id="d0e451" name="x-tspan" xmrk:locate_id="1">
   <segment>
     <source>Neck</source>
   </segment>
</unit>
<unit id="d0e457" name="x-tspan" xmrk:locate_id="2">
   <segment>
     <source>Body</source>
   </segment>
</unit>
<unit id="d0e463" name="x-tspan" xmrk:locate_id="3">
   <segment>
     <source>Electronics consist of</source>
   </segment>
</unit>
```

```
<unit id="d0e466" name="x-tspan" xmrk:locate_id="4">
  <segment>
    <source>three single-coil </source>
  </segment></unit>
<unit id="d0e469" name="x-tspan" xmrk:locate_id="5">
   <segment>
    <source>pick-ups, which can </source>
   </segment></unit>
<unit id="d0e472" name="x-tspan" xmrk:locate_id="6">
   <segment>
    <source>be selected using a </source>
   </segment>
</unit>
<unit id="d0e475" name="x-tspan" xmrk:locate_id="7">
   <segment>
    <source>rocker switch.</source>
   </segment>
</unit>
```

4. Translate the XLIFF file.

> ### Example 7.3 – XLIFF file from Example 7.2 translated into German
>
> ```
> <unit id="d0e445" name="x-tspan" xmrk:locate_id="0">
>     <source>Head</source>
>     <target>Kopf</target>
> </unit>
> <unit id="d0e451" name="x-tspan" xmrk:locate_id="1">
>     <source>Neck</source>
>     <target>Hals</target>
> </unit>
> <unit id="d0e457" name="x-tspan" xmrk:locate_id="2">
>     <source>Body</source>
>     <target>Korpus</target>
> </unit>
> <unit id="d0e463" name="x-tspan" xmrk:locate_id="3">
>     <source>Electronics consist of</source>
>     <target>Die Elektronik besteht aus </target>
> </unit>
> <unit id="d0e466" name="x-tspan" xmrk:locate_id="4">
> ```

```
<segment>
  <source>three single-coil </source>
  <target>drei Single-Coil-Tonabnehmern, </target>
</segment></unit>
<unit id="d0e469" name="x-tspan" xmrk:locate_id="5">
  <segment>
    <source>pickups, which can </source>
    <target> die über einen </target>
  </segment></unit>
<unit id="d0e472" name="x-tspan" xmrk:locate_id="6">
  <segment>
    <source>be selected using a </source>
    <target> Kippschalter angewählt </target>
  </segment>
</unit>
<unit id="d0e475" name="x-tspan" xmrk:locate_id="7">
  <segment>
    <source>rocker switch.</source>
    <target> werden können </target>
  </segment>
</unit>
```

5. Transform the translated XLIFF file into an SVG file (Figure 7.4).

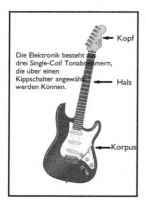

Figure 7.4 – German language SVG file

In the translated example (Figure 7.4), the longer German words run into the graphic. This problem can be solved in the original design of the graphic by allowing white space for text expansion. Or with SVG 1.2 there are advanced controls for text box constraints. These controls can be used to manage text expansion automatically.

CHAPTER 8
Translating Software User Interfaces

Translating software user interface text is a core competency for the XLIFF workflow. In most cases, the challenge is to extract the translatable text while preserving functionality. XLIFF 2.0 is optimized to provide for this, and you can find several commercial and open source tools that handle this workflow. We do not endorse any particular tool.

Table 8.1 shows some common programming languages that have translatable user interfaces along with the platform and method or file type used to store text strings.

Table 8.1 – Programming languages with translatable interfaces

Language	Platform	Text Resources Storage
C, C++	Windows, Linux, Mac OS X	■ .rc files on Windows ■ .po files on Linux
Objective-C, Objective-C++	Mac OS X, iOS	■ .po ■ .string
Java	Cross-platform	■ ListResourceBundles ■ PropertyResourceBundle ■ XML resources
C#, VisualBasic	Windows (.Net Framework)	■ ResX ■ XAML
JavaScript	Web Applications	■ In source code ■ JSon ■ XUL
PHP	Web Applications	In source code

PO (Portable Objects)

Most open source projects use GNU **gettext** to handle translatable user interface text. **gettext** supports localization of C, C++, Python, and other programming languages by extracting translatable text from source code and placing it in bilingual files that use the .po extension (see Example 8.1).

Example 8.1 – Sample .po file

```
# Translator Comment
#. Extracted Comment
#: myfile.c:12
#, flag
msgid "Original String 1"
msgstr "Translated String 1"

# Translator Comment
#. Extracted Comment
#: myfile.c:23
#, flag
msgid "Original String 2"
msgstr "Translated String 2"
```

The official manual[1] for the GNU **gettext** is available online from the GNU project. The XLIFF TC published official guidelines[2] for converting PO files to XLIFF 1.2.

Java

The Java object-oriented programming language lets you access translatable text using Java Resource Bundles, which are Java .properties files that contain locale-specific data. Because locale-sensitive objects, such as text strings, are extracted into .properties files, instead of being hardcoded into the program, you can support multiple language versions without rebuilding your application for each language.

You can transform resource bundles into XLIFF and translate them separately from of the code. The XLIFF TC published a reference guide for using Java Bundles and XLIFF,[3] which contains detailed information on how to use XLIFF with Java.

---

[1] http://www.gnu.org/software/gettext/manual/gettext.html
[2] http://docs.oasis-open.org/xliff/v1.2/xliff-profile-po/xliff-profile-po-1.2.html
[3] http://docs.oasis-open.org/xliff/v1.2/xliff-profile-java/xliff-profile-java-v1.2-cd02.html

Apple software

Apple software developers can find support for XLIFF in the Xcode development tool.[4]

Xcode uses resource files called *strings files*, which contain localizable strings.The Xcode user interface lets developers export the project's strings to an XLIFF file. After the XLIFF file is translated to the target language, the developer imports the XLIFF using the import function in Xcode. Xcode will decode the translated strings files from the XLIFF file and add the translations to the project folder, replacing any existing strings files.

Microsoft .Net Framework

Microsoft's Visual Studio IDE (Integrated Development Environment), which is used for developing Windows applications, provides the Multilingual App Toolkit[5] to export translatable text from their application into XLIFF files. When the translator completes the translation, the developer can import the translation into the program from Visual Studio.

JavaScript

There are several open source utilities available that claim to extract JavaScript to XLIFF. We do not endorse any particular utility. Transforming JavaScript to XLIFF can be straightforward. For example, consider the JavaScript in Example 8.2.

Example 8.2 – JavaScript sample

```
<script>
 function myUpdateFunction() {
    document.getElementById("intro").innerHTML = "Update complete.";
 }
</script>
```

---

[4] https://developer.apple.com/xcode/
[5] https://dev.windows.com/en-us/develop/multilingual-app-toolkit

When extracted for translation, it can be represented in XLIFF as shown in Example 8.3:

Example 8.3 – XLIFF file for Example 8.2

```
<unit id="u1">
  <segment id="s1">
    <source><pc id="1">Update complete.</pc></source>
  </segment>
</unit>
```

However, as with other languages, JavaScript provides certain methods that can make translation difficult. For example, functions such as toLowerCase() or toUpperCase() require special care because upper and lower case are irrelevant in some languages and treated differently in others.

Web applications

Here are a few of the more common technologies used for web applications:

- **Java Servlet and JavaServer pages:** These applications are written in Java and deployed using special web servers, such as Apache Tomcat[6] and GlassFish.[7] Because they use the Java language, localization follows the same workflow as Java.
- **AJAX:** These applications are written using several different languages – mainly Java and Python – and are compiled into JavaScript programs, which are deployed to standard web servers. You can localize either the source language or the JavaScript.
- **PHP:** This popular scripting framework stores translatable text in the source code. There are several open source utilities available that claim to extract PHP to XLIFF. We do not endorse any particular utility.

---

[6] http://tomcat.apache.org
[7] http://glassfish.org

XLIFF Core

The XLIFF 2.0 Core specifies the minimum set of features and functions required to perform a functional translation from a single source language to a single target language. This part of the book discusses these core functions, which include the following:

- Managing source format and hierarchy information for merging
- Preserving the source document format
- Presenting source text that is to be translated
- Providing places for the translated text to be managed
- Marking text that should or should not be translated
- Managing metadata that informs the translation process, such as the following:
  - Source language
  - Target language
  - Requirements around processing space
- Providing a mechanism for grouping and segmenting, including the following:
  - Nesting and structuring
  - Marking segments as eligible or not eligible for segmenting and re-segmenting
  - Changing the order of segments
- Providing for the following capabilities:
  - Sub-flows
  - Annotations
  - Bidirectional text
  - Fragment Identification
  - Extensibility

CHAPTER 9
Preserving Document Structure

To recreate the structure of the original file, merger agents need information about the structure of that file. XLIFF supports several methods for doing this. The two most frequently used methods are the minimalist method, which uses the `<skeleton>` element, and the maximalist method, which uses nested `<group>` elements.

XLIFF provides the elements shown in Table 9.1 to support these methods.

Table 9.1 – Elements that define the format and hierarchy of the content

Element	Purpose
`<xliff>`	root element for XLIFF documents
`<file>`	container for each source document
`<skeleton>`	container for source file's non-translatable structure
`<group>`	container for organizing units into a structured hierarchy
`<unit>`	static container for segments that can optionally be reordered
`<segment>`	container for the `<source>` element and `<target>` element that make up the segment
`<data>`	container for inline markup in the source text that should be preserved but should not be translated
`<originalData>`	wrapper element for one or more `<data>` elements

The minimalist method

A *skeleton* provides one method for preserving metadata and structure from a source file. This method is known as the *minimalist method* because a minimal amount of structural information is stored using the XLIFF standard, and most of the structural information is stored in a skeleton, which has an implementation defined structure.

Skeletons may be internal or external to your XLIFF file. In either case, you use the <skeleton> element, which, if present, must be the first child of the XLIFF <file> element.

In XLIFF 2.0, you either embed the skeleton data in the XLIFF file inside the <skeleton> element or use the href attribute on the <skeleton> element to link to an external file, which contains the skeleton information.

The contents of the skeleton are implementation specific and are not defined in the XLIFF standard. In our first example, we use a custom XML namespace with the prefix xmrk to markup information inside the skeleton, and in our second example, we simply embed RTF inside a single element in the xmrk: namespace. In both cases, the XLIFF standard does not define how to construct or reconstruct the file, it simply supplies a hook, <skeleton>, that implementations can use to store the information they need. Appendix B contains sample XSLT stylesheets to generate XLIFF with an internal skeleton and recreate a translated file using that skeleton.

You can use a skeleton for both XML and non-XML content.

Using an internal skeleton with XML source

If the source is XML and if storing it as XML is important, you can store the skeleton as XML. To store a skeleton that contains well-formed XML inside the XLIFF file, we declare our custom namespace (xmrk) at the top of the XLIFF file (see Example 9.1) and then use that namespace to capture the structural information.

Example 9.1 – Namespace declaration for xmrk:

```
<xliff xmlns="urn:oasis:names:tc:xliff:document:2.0"
       xmlns:xmrk="urn:xmarker"

    ...
```

The next step is to preserve the format and metadata of the source document to enable reconstruction of the target document after completing the translation. In Example 9.2, each XML element in the source XML file is represented by an element in the xmrk namespace that has the same name prefixed with xmrk. The hierarchy and structure of the original file are preserved, as are any attributes and attribute values. Each xmrk prefixed element contains an attribute, xmarker_idref, that points to the ID of the XLIFF <unit> element that contains the translatable text from that element.

For example, in Example 9.2, the element <xmrk:title xmarker_idref="title-2"/> on line 5 refers to a <title> element in the original document that is the child of the first <section> element in that document, and the value of the attribute xmarker_idref refers to the id attribute of the <unit> element in the XLIFF file that contains the text of the title. This skeleton contains enough information to reconstruct the original XML file structure.

Example 9.2 – Skeleton file using the xmrk: namespace

```
 1    <file>
 2     <skeleton>
 3      <xmrk:document xmarker_idref="document-0" status="draft" id="d1">
 4        <xmrk:section xmarker_idref="section-1" id="s1">
 5         <xmrk:title xmarker_idref="title-2"/>
 6         <xmrk:paragraph xmarker_idref="paragraph-3"/>
 7         <xmrk:section xmarker_idref="section-4" id="s1a">
 8          <xmrk:title xmarker_idref="title-5"/>
 9          <xmrk:paragraph xmarker_idref="paragraph-6"/>
10        </xmrk:section>
11        <xmrk:section xmarker_idref="section-7" id="s1b">
12          <xmrk:title xmarker_idref="title-8"/>
13          <xmrk:paragraph xmarker_idref="paragraph-9"/>
14        </xmrk:section>
15        <xmrk:section xmarker_idref="section-10" id="s1c">
16          <xmrk:title xmarker_idref="title-11"/>
17          <xmrk:paragraph xmarker_idref="paragraph-12"/>
18        </xmrk:section>
19       </xmrk:section>
20      </xmrk:document>
21     </skeleton>
22
23         . . .
```

Using an internal skeleton with non-XML source

If your source is not XML, you can embed structure information in the XLIFF file inside a single custom element. Be aware that if you embed the skeleton inside the XLIFF file, you must make sure the contents are valid XML. Thus, if there are characters, such as "<", which would be interpreted as part of XML markup, they must be escaped, either by using entities such as < or by placing the content in a CDATA section.

In Example 9.3, RTF code is inserted directly into a single custom element, <xmrk:rtf>. The processor must be able to read and interpret the RTF content to reconstruct the original file. How it does this is implementation specific and not part of the XLIFF standard.

Example 9.3 – Skeleton containing RTF structure information

```
<skeleton>

    <xmrk:rtf xmarker_idref="rtf file" status="draft" id="d1">

{\rtlch\fcs1 \af0 \ltrch\fcs0 \lang1024\langfe1024\noproof\insrsid14170
\charrsid14170 \hich\af31502\dbch\af31501\loch\f31502 History}
{\rtlch\fcs1 \af0\ltrch\fcs0 \lang1024\langfe1024\noproof\insrsid14170\par }
\pard\plain \ltrpar\ql \li0\ri0\sa200\sl276\slmult1\widctlpar
\wrapdefault\aspalpha\aspnum\faauto\adjustright\rin0\lin0\itap0 \rtlch
\fcs1 \af0\afs22\alang1025 \ltrch\fcs0 \f31506\fs22\lang1033\langfe1033
\cgrid\langnp1033\langfenp1033
{\rtlch\fcs1 \af0
\ltrch\fcs0 \lang1024\langfe1024\noproof\insrsid14170 This is a }
{\rtlch\fcs1 \af0 \ltrch\fcs0 \b\lang1024\langfe1024\noproof
\insrsid14170\charrsid14170 good}
{\rtlch\fcs1 \af0 \ltrch\fcs0 \lang1024\langfe1024\noproof
\insrsid14170  book.\par }

    </xmrk:rtf>

<skeleton>
```

The maximalist method

When a source document is not very complex and the document's structure could help the translator understand the context, it can be useful to preserve structural information and metadata in the XLIFF file, rather than relegating it to an external, implementation-specific format. This method is know as the *maximalist method* because it strives to preserve as much metadata and structure as possible using markup that is defined by the XLIFF standard. The maximalist method uses `<group>` elements and does not use a skeleton.

 The maximalist method works best for source documents that have few attributes and a simple structure.

Consider the XML file in Example 9.4.

Example 9.4 – Sample XML file for maximalist method

```
<document status="draft" id="d1">
  <section id="s1">
    <title>Birds in Oregon</title>
    <paragraph>Oregon is a mostly temperate state. There are many
    different kinds of birds that thrive</paragraph>
    <section id="s1a">
      <title>High Altitude Birds</title>
      <paragraph>Birds that thrive in the high altitude include the
      White-tailed Ptarmigan, Sharp-tailed Grouse, Yellow-billed
      Loon, Cattle Egret, Gyrfalcon, Snowy Owl, Yellow-billed
      Cuckoo, and Boreal Owl.</paragraph>
    </section>
    <section id="s1b">
      <title>Ocean Birds</title>
      <paragraph>Common ocean birds are Brandt's Cormorant,
      Double-crested Cormorant, Pelagic Cormorant, Pigeon
      Guillemot, and the Tufted Puffin.</paragraph>
    </section>
    <section id="s1c">
      <title>Desert Birds</title>
      <paragraph>Birds you find in the desert include the Sage
      Grouse, California Quail, and Prairie Falcon.</paragraph>
    </section>
  </section>
</document>
```

You use XLIFF <group> elements to capture structural elements such as <document> and <section>. The <group> elements are nested to parallel the structure, and each <group> element contains a name attribute, which captures the name of the original element. You use the XLIFF <unit> element to capture the contents of elements that contain translatable text, just as you would in any translation. However, you add the name attribute to the <unit> element to capture the name of the original element.

You preserve attribute names and values using the <metadata> module. In Example 9.5, the elements using the prefix mda contain the metadata. See Chapter 19 for more information about the metadata module.

 You can also use XML processing instructions to preserve attribute names. However, the XLIFF specification does not require processing instructions to be preserved in a round trip, so they could be removed during processing.

Example 9.5 shows the XLIFF 2.0 file created for the source file in Example 9.4. For many XML files, this method captures enough information to reconstruct the original file structure after translation, but with more complex files, the markup can become hard to manage.

Example 9.5 – XLIFF file for Example 9.4 using maximalist method

```
<xliff xmlns="urn:oasis:names:tc:xliff:document:2.0"
       xmlns:fs="urn:oasis:names:tc:xliff:fs:2.0"
       xmlns:mda="urn:oasis:names:tc:xliff:metadata:2.0"
       version="2.0"
       srcLang="en">
  <file id="groups">
    <group id="N65541xdocument" name="document">
      <mda:metadata>
        <mda:metaGroup category="XMLattribute">
          <mda:meta type="status">draft</mda:meta>
          <mda:meta type="id">d1</mda:meta>
        </mda:metaGroup>
      </mda:metadata>
      <group id="N65541bxmarksection-1" name="section">
        <mda:metadata>
          <mda:metaGroup category="XMLattribute">
            <mda:meta type="id">s1</mda:meta>
          </mda:metaGroup>
        </mda:metadata>
        <unit id="title-2" name="title">
          <segment>
            <source>Birds in Oregon</source>
```

```
  </unit>
  <unit id="paragraph-3" name="paragraph">
    <segment>
      <source>Oregon is a mostly temperate state. There are
      many different kinds of birds that thrive</source>
    </segment>
  </unit>
  <group id="N65547bxmarksection-4" name="section">
    <mda:metadata>
      <mda:metaGroup category="XMLattribute">
        <mda:meta type="id">s1a</mda:meta>
      </mda:metaGroup>
    </mda:metadata>
    <unit id="title-5" name="title">
      <segment>
        <source>High Altitude Birds</source>
      </segment>
    </unit>
    <unit id="paragraph-6" name="paragraph">
      <segment>
        <source>Birds that thrive in the high altitude
        include the Whitetailed Ptarmigan, Sharp-tailed
        Grouse, Yellow-billed Loon, Cattle Egret, Gyrfalcon,
        Snowy Owl, Yellow-billed Cuckoo, and Boreal
        Owl.</source>
      </segment>
    </unit>
  </group>
  <group id="N65553bxmarksection-7" name="section">
    <mda:metadata>
      <mda:metaGroup category="XMLattribute">
        <mda:meta type="id">s1b</mda:meta>
      </mda:metaGroup>
    </mda:metadata>
    <unit id="title-8" name="title">
      <segment>
        <source>Ocean Birds</source>
      </segment>
    </unit>
    <unit id="paragraph-9">
      <segment>
        <source>Common ocean birds are Brandt's Cormorant,
        Double-crested Cormorant, Pelagic Cormorant, Pigeon
        Guillemot, and the Tufted Puffin.</source>
      </segment>
    </unit>
  </group>
  <group id="N65559bxmarksection-10" name="section">
    <mda:metadata>
```

```
              <mda:metaGroup category="XMLattribute">
                <mda:meta type="id">s1c</mda:meta>
              </mda:metaGroup>
            </mda:metadata>
            <unit id="title-11" name="title">
              <segment>
                <source>Desert Birds</source>
              </segment>
            </unit>
            <unit id="paragraph-12" name="paragraph">
              <segment>
                <source>Birds you find in the desert include the Sage
                Grouse, California Quail, and Prairie
                Falcon.</source>
              </segment>
            </unit>
          </group>
        </group>
      </group>
    </file>
  </xliff>
```

CHAPTER 10
Marking up Text for Translation

XLIFF 2.0 files combine text to be translated, information about the translation job, information about the source structure, and after translation, translated text. Table 10.1 shows the elements used to mark up text in XLIFF files.

Table 10.1 – XLIFF 2.0 text elements

Element	Purpose
Container elements	
`<source>`	Container for source text to be translated
`<target>`	Container for translated text
Inline elements	
`<cp>`	Marks up Unicode characters that are invalid in XML
`<ph>`	Marks up a standalone element, for example, an HTML ` ` element
`<pc>`	Marks up a well-formed span that has inline markup
`<sc>`	Marks the beginning of a span that cannot be represented by `<ph>` because it begins and ends in different segments or would not be well-formed.
`<ec>`	Marks the end of span that began with `<sc>`
`<mrk>`	Attaches an annotation to a well-formed span
`<sm>`	Marks the beginning of an annotation that cannot be represented by `<mrk>` because it begins and ends in different segments or would not be well-formed.
`<em>`	Marks the end of an annotation that began with `<sm>`.

Examples of XLIFF inline markup

The following examples show how to markup inline elements. The inline elements are the same for either container element: `<source>` or `<target>`. The only difference is the container.

Example 10.1 – Example with no inline elements

For the following source content:

```
<p>Why is the guitar such a popular musical instrument?</p>
```

One valid way to extract the XLIFF 2.0 markup is the following:

```
<segment>
  <source>Why is the guitar such a popular musical instrument?</source>
</segment>
```

Adding the translation to the XLIFF file gives you the following:

```
<segment>
  <source>Why is the guitar such a popular musical instrument?</source>
  <target>Warum ist die Gitarre ein so beliebtes Musikinstrument?</target>
</segment>
```

Example 10.2 – Example with well-formed inline elements in one segment

For the following source content:

```
<p>Why is the <b>guitar</b>such a popular musical instrument?</p>
```

Here is the XLIFF 2.0 file, including the translated text. This example uses the `<pc>` element:

```
<segment>
  <source>Why is the
  <pc type="fmt" subType="xlf:b"
      id="pc1">guitar</pc>such a popular musical instrument?</source>
  <target>Warum ist die
  <pc type="fmt" subType="xlf:b"
      id="pc1">Gitarre</pc>ein so beliebtes Musikinstrument?</target>
</segment>
```

In Example 10.2, the `<pc>` element takes the type attribute, which defines the type of markup (in this case fmt for formatting). It also takes the subType attribute, which identifies the type of formatting (in this case, `<xlf:b>` for bold). Finally, it takes the id attribute, which links the `<pc>` element in the `<source>` with the parallel element in the `<target>`.

Example 10.3 – Example with malformed inline elements in one segment

For the following source content:

```
<p>Why is the <b><i>guitar </b>such a popular</i> musical instrument?</p>
```

Here is the XLIFF 2.0 markup for the source text, using the `<sc>` and `<ec>` elements.

```
<segment>
  <source>Why is the
  <sc id="s1" type="fmt" subType="xlf:b" />
  <sc id="s2" type="fmt" subType="xlf:i" />guitar
  <ec startRef="s1" type="fmt" subType="xlf:b" />such a popular
  <ec startRef="s2" type="fmt" subType="xlf:i" />musical instrument?</source>
</segment>
```

In Example 10.3, the `<b>` and `<i>` elements in the source file are not well-formed XML. Therefore, we can't use the `<pc>` element we used in Example 10.2. Instead, we use the `<sc>` and `<ec>` elements, which can capture badly formed XML. The only new attribute in this example is startRef, which is used on `<ec>` to point to the id attribute on `<sc>`.

Example 10.4 – Example with well-formed inline elements that cross segments

For the following source content:

```
<p>Why is the guitar such a popular <b>musical instrument? Musical
instruments</b> are common here.</p>
```

Here is the XLIFF 2.0 markup using the `<sc>` and `<ec>` elements.

```
<segment>
  <source>Why is the guitar such a popular
    <sc id="s1" type="fmt" subType="xlf:b"/>musical instrument?
  </source>
</segment>
<segment>
  <source>Musical instruments<ec startRef="s1" type="fmt" subType="xlf:b"/>
    are common here.
  </source>
</segment>
```

In Example 10.4, you can see that the `<sc>` and `<ec>` elements work the same way across segments as they do within a single segment.

Identifying a segment as having been translated

Often, you need to track what stage various segments are in with respect to the translation workflow. You use the `state` attribute on the `<segment>` element to do this. The `state` attribute can take the values: `initial`, `translated`, `reviewed`, or `final`. Example 10.5 shows the `state` on a segment.

Example 10.5 – Using the `state` attribute

```
<segment state="translated">
  <source>Why is the guitar such a popular musical instrument?
  </source>
  <target>Warum ist die Gitarre ein so beliebtes Musikinstrument?
  </target>
</segment>
```

Marking up untranslatable text

Not all text should be translated. The `translate` attribute lets you identify content – a word, a phrase, a paragraph, a section, or even a whole document – that should not be translated. The `translate` attribute has two possible values, `yes` and `no`. XLIFF 2.0 lets you use the translate attribute on any of the elements shown in Table 10.2.

Table 10.2 – Elements that allow the `translate` attribute

Element	Scope
`<file>`	indicates that an entire file should or should not be translated
`<group>`	indicates that a section should or should not be translated
`<unit>`	indicates that a paragraph should or should not be translated
`<mrk>`	indicates that a sentence or fragment should or should not be translated
`<sm>`	indicates that a sentence or fragment should or should not be translated

Although you can use the `translation` attribute on the inline elements `<mrk>` and `<sm>`, you take a risk because you may need to translate the contents of an inline element in one target language but not in another.

The default value of the `translate` attribute is `yes`. However, when the value of the `translate` attribute is set one way on an element and another way on that element's child, the value in the child element overrides the value on the parent for that child.

Example 10.6 shows how the translate attribute works. In that example, the strings "Crawl," "Run," and "Fly" will be translated, but the string "Walk" will not.

Example 10.6 – Scope of the `translate` attribute

```
<file translate="no" id="f1">
  <group id="g1" translate="yes">
    <unit translate="no" id="u1">
      <segment id="s1">
       <source><mrk translate="yes" type="term" id="m1">Crawl</mrk></source>
      </segment>
    </unit>
  </group>
  <group id="g2">
    <unit id="u2">
      <segment id="s2">
        <source><mrk id="m2" type="term">Walk</mrk></source>
      </segment>
    </unit>
  </group>
  <group id="g3">
    <unit id="u3">
      <segment id="s3">
        <source><mrk translate="yes" type="term" id="m3">Run</mrk></source>
      </segment>
    </unit>
  </group>
  <group id="g4">
    <unit translate="yes" id="u4">
      <segment id="s4">
        <source><mrk id="m4" type="term">Fly</mrk></source>
      </segment>
    </unit>
  </group>
</file>
```

Translation metadata

The XLIFF standard allows you to add information to an XLIFF file about the content being translated – both the source and target – and about the translation workflow. You can add metadata as XML attributes or as content in XML elements, and you can add and update metadata throughout the translation workflow.

 The metadata described in this section addresses only the XLIFF 2.0 Core. Metadata for XLIFF 2.0 Modules is discussed in Part IV, "XLIFF Modules."

XLIFF metadata held in attributes

The attributes in Table 11.1 hold information applicable to the translation workflow. The table describes each attribute and shows which elements each attribute can be used on.

Table 11.1 – XLIFF metadata attributes

Attribute	Element	Value(s)	Description
srcLang	<xliff>	A language code conforming to BCP 47[a]	The language of the source document.
trgLang	<xliff>	A language code conforming to BCP 47	The language of the translation.
translate	<file>	One of: yes, no	Whether or not the source text in the scope of the translate attribute should be translated.
order	<target>	A positive integer	When the order of content in a segment differs between source and target, the order attribute defines the desired order (see Example 12.1).

Attribute	Element	Value(s)	Description
state		One of: initial, translated, reviewed, final	The translation state of a segment.
subState		Prefix and a sub-value separated by a colon (:). The prefix uniquely identifies the user, and the sub-value is a user-defined string. The prefix xlf is reserved for the XLIFF specification. Users may define other prefixes and sub-values.	User-defined state of the translation of a segment.
copyOf	<ph>	NMTOKEN[b]	Refers to the id of another element that this element is a copy of.
canReorder	<ph>	yes if the content can be re-ordered, firstNo when the content is the first element in a sequence and cannot be reordered, and no when the content is another element of such a sequence.	Information about what a translator may do with segments — canReorder set to no means the element may not be re-ordered.
canResegment	<file>	One of: yes, no	Information about what a translator may do with segments — canResegment set to no means the element may not be re-segmented.

Attribute	Element	Value(s)	Description
canCopy	`<file>`	One of: yes, no	Information about what a translator may do with segments — `<canCopyattribute>` set to no means the `<segment>` element may not be copied.
canDelete	`<file>`	One of: yes, no	Information about what a translator may do with segments — `canDelete` set to no means the `<segment>` element may not be deleted.
canOverlap	`<file>`	One of: yes, no	Information about what a translator may do with segments — `canOverlap` set to no means the `<segment>` element may not be over-lapped. This helps prevent situations where you end up with an overlap within a segment that includes a start code without a corresponding end code or an end code without a corresponding start code.

[a] http://tools.ietf.org/html/bcp47BCP 47 (Best Current Practice 47) defines best practices for identifying a language. XLIFF uses BCP 47 for this attribute and trgLang.

[b] A token that contains only letters, digits, periods, hyphens, underscores or colons. A token can start with any of these characters.

The attributes in Table 11.2 contain information about the translation workflow, but they may also contain additional information that applies to the content.

Table 11.2 – Metadata attributes that include data

Attribute	Element	Value(s)	Description
dir	`<data>`	Text direction	`ltr` (Left-To-Right), `rtl` (Right-To-Left), or `auto`.[a]
srcDir	`<file>`	Text direction of the source text	`ltr` (Left-To-Right), `rtl` (Right-To-Left), or `auto`.
trgDir	`<file>`	Text direction of the target text	`ltr` (Left-To-Right), `rtl` (Right-To-Left), or `auto`.
subFlows	`<ph>`	NMTOKEN	Information about embedded, often translatable, text to help recompose sub-flows when they are contained in separate elements.
subFlowsEnd	`<pc>`	NMTOKEN	The end of a sub-flow.
subFlowsStart	`<pc>`	NMTOKEN	The start of a sub-flow.
hex	`<cp>`	A hexadecimal code point. Two hexadecimal digits to represent each octet of the Unicode code point.	Used to represent valid Unicode characters that invalid in XML.
isolated	`<sc>`, `<ec>`	One of: yes, no	If `yes`, the corresponding start or end marker is in a different `<unit>`. If `no`, they are in the same `<unit>`.
startRef	`<ec>`, `<em>`	NMTOKEN	Reference to the `id` attribute of the corresponding `<sc>` or `<sm>` element.

[a] The value auto means the text direction is determined heuristically, based on the first strong directional character in scope, see http://www.unicode.org/reports/tr9/.

XLIFF metadata held in elements

While the majority of XLIFF 2.0 elements contain structural information or content, the elements in Table 11.3 can contain metadata about the translation.

Table 11.3 – Metadata elements in XLIFF

Element	Purpose
`<notes>`	Organizes notes.
`<note>`	Contains end-user readable comments and annotations. A note can be applied to: `<source>`, `<target>`, `<unit>`, `<group>`, or `<file>`
`<originalData>`	Container inside a `<unit>` that wraps one or more `<data>` elements. A unit-level collection of original data for the inline code elements, which uses `<data>` elements to represent the code.
`<data>`	Contains the original content from an inline element in the source. For example, the HTML ` ` element could be represented as: `<data id="d2">&lt;br/></data>`
`<ignorable>`	Contains extracted content not in a segment and not to be translated.
`<mrk>`	Contains an annotation for a marked span – the text between the mark start and end tags is the span. E.g., a glossary term could be annotated as follows: `<mrk id="m1" type="term" ref="#g1">bill.</mrk>` This could resolve to a Glossary Entry such as: ```<gls:glossEntry id="g1" ref="#m1"` ` xmrk:legal_approval_req="lax">` ` <gls:term>bill</gls:term>` ` <gls:translation>Schnabel</gls:translation>` ` <gls:definition source="dictionary.com">The hard` ` usually pointed parts that cover a bird's mouth.` ` </gls:definition>` `</gls:glossEntry>```
`<sm>`	Annotation start marker where the spanning marker cannot be used.
`<em>`	Annotation end marker where the spanning marker cannot be used.

CHAPTER 12

Segmentation, and Sub-flows

Controlling the granularity and structure of translatable text is an important aspect of the translation process. In Chapter 9, *Preserving Document Structure*, we saw two methods for preserving structure. This chapter shows how the granularity of text can be controlled using segmentation and how text that must be handled separately from the main flow of text can be managed with sub-flows.

Segmentation splits a portion of translatable text into smaller pieces called *segments*. It optimizes the size and delineation (the starting and finishing points) of text for translation. Segmentation also helps optimize the use of translation memory. Segmentation is generally automated but often needs to be adjusted by human translators.

Sub-flows are pieces of text, usually translatable, which are not part of the normal flow of text in the document being translated. Examples include text for a bookmark or text displayed as a mouse-over for an image.

Segmentation

Translation Memory (TM) engines usually store sentences. Therefore, to increase the probability of finding matches in the TM system, paragraphs of text are usually segmented into sentences.

With segmentation, a discreet set of text elements (such as titles, paragraphs, lists, methods, classes, etc.) is marked as a managed unit. (In this context, unit is meant generically, not as the XLIFF <unit> element). The format for segmenting and re-segmenting in XLIFF 2.0 was designed to address user dissatisfaction with XLIFF 1.2 segmentation. As a result, segmentation in XLIFF 2.0 has been improved and is significantly different from segmentation in XLIFF 1.2.

Segmentation can happen in two different stages of the XLIFF workflow:

- When you extract text for translation, but before you create the XLIFF file
- After you create the XLIFF file

The elements in Table 12.1 are used to define segments.

Table 12.1 – Segmentation elements

Element	Purpose
`<segment>`	Container for translatable source and target text. Segments may be reordered.
`<ignorable>`	Container for extracted content that is not included in a segment and is not translated. Use `<ignorable>` to store white space or code between segments.
`<source>`	Container for text to be translated, including any inline elements that mark up flows
`<target>`	Container for translated text, including any inline elements that mark up flows
`<mrk>`	Marks spans of text that may contain annotations that affect grouping and segmenting

Each managed unit of text in the document is divided into `<segment>` elements. Each `<unit>` can contain one or more `<segment>` elements. Each `<segment>` element has exactly one `<source>` element, which contains the source content, and one optional `<target>` element, which can be empty or can contain the translation of the source. The `<unit>` element may also contain one or more `<ignorable>` elements.

Example 12.1 shows some familiar text divided into segments. In most forms of English, the following would be a reasonable cadence: "You have become powerful; I sense in you the dark side." However, the character of Yoda in the Star Wars movies uses a different cadence: "Powerful you have become; the dark side I sense in you."

Example 12.1 – Segmentation in XLIFF 2.0

```
<unit id="show_order">
  <segment id="s1">
    <source xml:lang="en-US">You have become</source>
  </segment>
  <segment id="s2">
    <source xml:lang="en-US">powerful;</source>
  </segment>
  <segment id="s3">
    <source xml:lang="en-US">I sense in you</source>
  </segment>
  <segment id="s4">
    <source xml:lang="en-US">the dark side</source>
  </segment>
</unit>
```

If you want to re-segment a unit to accommodate Yoda-English, you would use the order attribute on the <target> elements, as shown in Example 12.2.

Example 12.2 – Segment ordering

```
<unit id="show_order">
  <segment id="s1">
    <source xml:lang="en-US">You have become</source>
    <target xml:lang="en-x-YODA" order="2">you have become; </target>
  </segment>
  <segment id="s2">
    <source xml:lang="en-US">powerful;</source>
    <target xml:lang="en-x-YODA" order="1">Powerful </target>
  </segment>
  <segment id="s3">
    <source xml:lang="en-US">I sense in you</source>
    <target xml:lang="en-x-YODA" order="4">I sense in you</target>
  </segment>
  <segment id="s4">
    <source xml:lang="en-US">the dark side.</source>
    <target xml:lang="en-x-YODA" order="3">the dark side </target>
  </segment>
</unit>
```

In a translation workflow, you may need to change the level of granularity of a segment to improve translation efficiency or leverage translation memory more effectively.

Consider the following scenario. An extractor agent segments the initial XLIFF file into paragraph-level segments, but your translation memory uses sentence-level segments. XLIFF 2.0 lets you re-segment at the sentence level. However, you must follow certain processing requirements and constraints. We will start with Example 12.3, which is segmented at the paragraph level. Each segment contains multiple sentences.

Example 12.3 – Changing segment granularity

```
<unit id="u4">
 <segment state="initial" subState="xmrk:pre-leveraged">
   <source>The first song is about the sky. The second song is about
           the ocean. The third song is about the mountains.</source>
 </segment>
 <segment canResegment="no">
   <source>The album was released last year. It went straight to
           number one.</source>
 </segment>
</unit>
```

The goal is to re-segment each element at the sentence level. We can re-segment the first element, but we cannot re-segment the second because the canResegment attribute is set to no.

According to the XLIFF 2.0 specification, "All new or <ignorable> elements created and their <source> and <target> children MUST have the same attribute values as the original elements they were created from, as applicable, except for the id attributes and, possibly, for the order, state and subState attributes." Since the state and subState attributes exist in the original XLIFF file, we carry them forward during re-segmentation.

After re-segmenting, each sentence in the first segment is contained in a separate element, the attributes from the original segment have been copied into each new segment, and the second segment remains unchanged (see Example 12.4).

Example 12.4 – Re-segmented source text

```
<unit id="u4">
 <segment state="initial" subState="xmrk:pre-leveraged">
  <source>The first song is about the sky.</source>
 </segment>
 <segment state="initial" subState="xmrk:pre-leveraged">
  <source>The second song is about the ocean.</source>
 </segment>
 <segment state="initial" subState="xmrk:pre-leveraged">
  <source>The third song is about the mountains.</source>
 </segment>
 <segment canResegment="no">
  <source>The album was released last year. It went straight to number one.
  </source>
 </segment>
</unit>
```

Preserving content

The segmentation process must preserve all text that needs to be translated and must also preserve any formatting. The `<ignorable>` element identifies extracted content that is not included in a segment and that is not intended to be translated. Tools can use `<ignorable>` to store ignorable content such as markers for images and white space between two segments. Consider the HTML fragment in Example 12.5:

Example 12.5 – HTML fragment

```
<p>This picture is nice: <img src="cat.png" />. The cat looks
   terrific!</p>
```

This fragment could be extracted as shown in Example 12.6.

Example 12.6 – HTML fragment extracted into XLIFF

```
<unit id="u1">
  <segment id="s1">
    <source>This picture is nice: <ph id="1"/>. The cat looks
            terrific!</source>
  </segment>
</unit>
```

After segmentation, Example 12.6 would look like Example 12.7

Example 12.7 – HTML fragment segmented into sentences and ignorable sections

```
<unit id="1">
  <segment id="s1">
    <source>This picture is nice:</source>
  </segment>
  <ignorable id="i1">
    <source> <ph id="1"/>. </source>
  </ignorable>
  <segment id="s2">
    <source>The cat looks terrific!</source>
  </segment>
</unit>
```

Notice that the concatenation of the content from all <source> elements (including both and <ignorable> elements) matches the content of the original, unsegmented <source> element, including white space. You can preserve white space between sentences and inline elements in <ignorable> elements to ensure it doesn't get lost.

Sub-flows

Most of the time, translatable text in a source document is clearly accessible to the extractor agent and can be modeled using , <source>, and inline elements. In Example 12.8 and Example 12.9, the translatable text is easily accessible to an extractor agent.

Example 12.8 – RTF without sub-flows

```
\pard\plain \fs20 After the punk rock style had been
 established, different styles
 emerged with their own subcultures.
```

Example 12.9 – HTML without sub-flows

```
<p>Mick Jones renewed his interest in the Vintage
   <strong>'72 Telecaster Thinline </strong> guitar.
   <img src="smileface.png" />
</p>
```

However, the bookmark (`{\bkmkstart subcultures}`) in Example 12.10 is a sub-flow. It does not appear as part of the RTF paragraph, but it still must be translated.

Example 12.10 – RTF with bookmark

```
\pard\plain \fs20 After the punk rock style had been
 established, {\bkmkstart subcultures} different styles
 emerged with their own subcultures.{\bkmkend subcultures}
```

XLIFF 2.0 marks sub-flows using the `subFlows` attribute on the `<ph>` element. Tools can then reference sub-flows within units and extract them for translation. Example 12.11 shows how to handle a sub-flow.

Example 12.11 – XLIFF for Example 12.10 with a sub-flow for the bookmark

```
<unit id="sf1">
  <segment>
    <source>subcultures</source>
    <target>Subkulturen</target>
  </segment>
</unit>
<unit id="sf2">
  <segment>
    <source>subcultures</source>
    <target>Subkulturen</target>
  </segment>
</unit>
<unit id="sf3">
  <segment>
    <source>After the punk rock style had been established,
      <ph id="s1" subFlows="sf1" />different styles emerged
        with their own subcultures.
      <ph id="s2" subFlows="sf2" /></source>
    <target>Nachdem sich der Punk-Rock etabliert hatte,
      <ph id="s1" subFlows="sf1" />entstanden verschiedene
        Stilrichtungen mit eigenen Subkulturen.
      <ph id="s2" subFlows="sf2" /></target>
  </segment>
</unit>
```

Adding a title and alternate text to the HTML example adds sub-flows. These sub-flows are not displayed unless an event such as a mouse-over occurs. However, the text still must be translated. In the example below, the sub-flow is `alt="Fender Guitar"`

Example 12.12 – HTML with alt text

```
<p>Mick Jones renewed his interest in the Vintage
   <strong>'72 Telecaster Thinline </strong> guitar.
   <img src="smileface.png" title="Jones Happy Face" alt="Fender Guitar"/>

</p>
```

Example 12.13 shows how the <ph> elements, along with the subFlows attributes, present sub-flows to the translator. The <originalData> and <data> elements inside <unit id="p1"> capture the original structure and content from the file for later reconstruction. The sub-flow content is separated into two units, <unit id="ssf1"> and <unit id="ssf2">. The <unit id="p1"> element contains a segment with the regular text, plus <ph> elements that point to the <data> elements (for structure) and text units (for translatable text) for the sub-flows.

Example 12.13 – XLIFF for Example 12.12 with sub-flow for alt text

```
<unit id="ssf1">
  <segment>
    <source>Jones Happy Face</source>
  </segment>
</unit>
<unit id="ssf2">
  <segment>
    <source>Fender Guitar</source>
  </segment>
</unit>
<unit id="p1">
  <originalData>
    <data id="dr1">&lt;img src="smileface.png" title="Jones Happy Face"
         alt="Fender Guitar"/&gt;</data>
    <data id="dr22">&lt;br/&gt;</data>
  </originalData>
  <segment>
    <source>Mick Jones renewed his interest in the Vintage
    <pc id="1">'72 Telecaster Thinline</pc>guitar.
    <ph id="ph1" dataRef="dr22" />
    <ph id="ph2" subFlows="ssf1 ssf2" dataRef="dr1" /></source>
  </segment>
</unit>
```

CHAPTER 13

Annotations and Bidirectional Text

At different stages of the translation workflow, you may need to annotate the text. For example, you may discover new information about the translation such as words or phrases that should not be translated or hints that other resources exist to aid in the translation. XLIFF 2.0 defines the following types of annotations:

- Translate annotations
- Term annotations
- Comment annotations
- Custom annotations

Translate annotations

Translate annotations indicate whether a span of text is translatable or not translatable. This annotation overrides the currently active value of the translate attribute. In Example 13.1, the phrase *bona fide* is marked to not be translated although the surrounding sentence will be translated.

Example 13.1 – Translate annotation

```
<!-- Translate Annotation -->
<unit id="u6">
  <segment>
    <source>He is a <mrk id="mrk1" translate="no">bona fide</mrk> expert.
    </source>
  </segment>
</unit>
```

Term annotations

Term annotations identify a word or phrase as a term. You can add information, such as a glossary definition, to this annotation. In Example 13.2, the annotation links the term *bona fide* to a glossary definition, which is shown at the beginning of the unit.

Example 13.2 – Term annotation

```
<!-- Term Annotation -->
<unit id="u7">
  <gls:glossary>
    <gls:glossEntry id="gls1">
      <gls:term>bona fide</gls:term>
      <gls:definition source="http://www.merriam-webster.com">Neither
        specious nor counterfeit.</gls:definition>
    </gls:glossEntry>
  </gls:glossary>
  <segment>
    <source>He is a
    <mrk id="mrk1" type="term" ref="gls=gls1"
        translate="no">bona fide</mrk>expert.</source>
  </segment>
</unit>
```

 Term annotations are independent of the Glossary module. Also, it is valid to include the `translate` attribute on a term annotation, which we have done in this case, since *bona fide* is a Latin term that shouldn't be translated.

Comment annotations

Comment annotations associate a span of text with a comment, which can then be associated with a note.

Example 13.3 – Comment annotation

```
<!-- Comment Annotation -->
<unit id="u9">
  <notes>
    <note id="note1" appliesTo="target">It is okay to use the Latin
      phrase 'bona fide' in German. The German word 'anerkannter'
      is more generic.</note>
  </notes>
  <segment>
    <source>He is a bona fide expert.</source>
    <target>Er ist ein
      <mrk id="mrk1" type="comment" ref="#n=note1">anerkannter</mrk>
      Experte.</target>
  </segment>
</unit>
```

Custom annotations

Custom Annotations allow agents to add annotations that are not translation, term, or comment related.

The following example shows a custom annotation that uses an employee roster to augment the translation.

Example 13.4 – Custom annotation

```
<!-- Custom Annotation, value and ref are user-defined -->
<unit id="u10">
  <segment>
    <source><mrk id="mrk1" type="CompanyRoster:name"
                       value="Eric_Smart">He</mrk> is
                       a bona fide expert.
    </source>
  </segment>
</unit>
```

Bidirectional text

XLIFF 2.0 supports markup to identify the direction of text, but it does not prescribe a method for bidirectional processing. You can set source and target directionality at the file, group, or unit level using the `srcDir` and `trgDir` attributes. And you can set the direction of a text string within inline elements using the `<pc>`, `<sc>`, or `<ec>` elements.

XLIFF 2.0 complies with the rules for processing bidirectional text defined by the Unicode Consortium.[1]

---

[1] The rules are defined in the Unicode specification: http://www.unicode.org/reports/tr9/

CHAPTER 14

Fragment Identification

Fragment identifiers in XLIFF 2.0 serve two purposes. First, they provide a way for external tools to reference XLIFF elements such as `<file>`, `<group>`, `<unit>`, `<note>`, `<data>`, `<target>`, `<segment>`, and `<ignorable>` and elements from modules or extensions. Second, they provide a way to reference these elements from within an XLIFF file.

Why use fragment identifiers instead of XML IDs?

Standard XML IDs must be unique across an entire document. If an XML ID is duplicated, a reference to that ID will be ambiguous. To support distributed workflows, XLIFF documents must tolerate being split, combined, distributed among agents, and recombined. As a result, an XLIFF file may end up with duplicate IDs. The XLIFF 2.0 fragment identification feature provides a mechanism that supports duplicate IDs and a syntax for unambiguously referring to IDs.

Example 14.1 shows a typical XLIFF file (`bird.xlf`) with duplicate ID references (`id="u1"` on each of the two `<unit>` elements).

Example 14.1 – XLIFF file with duplicate ID references

```
<?xml version="1.0"?>
<xliff xmlns="urn:oasis:names:tc:xliff:document:2.0"
       version="2.0" srcLang="en">
 <file id="f1">
  <unit id="u1">
   <segment>
    <source>It was a unique bill.</source>
   </segment>
  </unit>
 </file>
 <file id="f2">
  <unit id="u1">
   <segment>
    <source>Many birds have unique bills.</source>
   </segment>
  </unit>
 </file>
</xliff>
```

It is perfectly legal to have two different <file> elements, each of which contains a <unit> element with an @id attribute that has the value u1.

A reader parsing the XLIFF file in Example 14.1 might look for the text of a <source> element that is a descendant of the <unit> element whose id attribute has the value u1. For example, in Java the expression might be written as shown in Example 14.2:

Example 14.2 – Java code to find an element by its id attribute

```
DocumentBuilder myXLIFF;
    try {
        myXLIFF = DocumentBuilderFactory.newInstance().newDocumentBuilder();
        Document XLIFFdocument = myXLIFF.parse(new File("/bird.xlf"));
        XPath findText = XPathFactory.newInstance().newXPath();
        String expression = "//unit[@id='u1']//source/text()";
    Node widgetNode = (Node) findText.evaluate(expression, XLIFFdocument,
                        XPathConstants.NODE);
        System.out.println("1. " + widgetNode);
    }
```

Another example is the mechanism used by HTML: <a href= "html_tips.htm#tips">, where tips is an ID somewhere in the file html_tips.htm. The XPATH standard would use this code: //unit[@id='u1']//source/text(). In both cases the result is ambiguous because the id attribute is not unique.

XLIFF 2.0 fragment identifiers avoid this problem. The IDs in an XLIFF file look like XML IDs, but the rules for duplication are relaxed. An XLIFF ID must be unique among its sibling elements of the same type. For example, in any <file> element, all child <unit> elements must have unique IDs. However, two <unit> elements may have duplicate IDs if they have different parent <file> elements. In Example 14.1, if the element <file id=f1"> had two <unit> elements as children, those two child elements could not have duplicate IDs. But since the two <unit> elements in Example 14.1 have different parent <file> elements, they may have duplicate IDs.

Fragment identifier syntax

The XLIFF 2.0 fragment identification feature defines a syntax that allows unambiguous ID references. The syntax builds a path to the ID starting either at a filename or at a relative position in the same file. Example 14.3 shows the syntax for fragment identification in EBNF. The fragment identifier may be preceded by the name of the file that contains the ID being referred to.

Example 14.3 – EBNF for the fragment identifier syntax

```
<expression>        ::= "#" ["/"] <selector> {<selectorSeparator> <selector>}
<selector>          ::= [<prefix> <prefixSeparator>] <id>
<prefix>            ::= NMTOKEN
<id>                ::= NMTOKEN
<prefixSeparator>   ::= "="
<selectorSeparator> ::= "/"
```

Table 14.1 shows the allowable values for the fragment identifier prefix. Other values of the prefix may be valid if you are using the extensibility feature of XLIFF.

Table 14.1 – Fragment identifier prefixes

Prefix	Definition
prefix f	a `<file>` id that has a value unique among all `<file>` id attribute values within the enclosing `<xliff>` element
prefix g	a `<group>` id that has a value unique among all `<group>` id attribute values within the enclosing `<file>` element
prefix u	a `<unit>` id that has a value unique among all `<unit>` id attribute values within the enclosing `<file>` element
prefix n	a `<note>` id that has a value unique among all `<note>` id attribute values within the immediate enclosing `<file>`, `<group>`, or `<unit>` element
prefix d	a `<data>` id that has a value unique among all `<data>` id attribute values within the enclosing `<unit>` element
prefix t	an id for an inline element in the `<target>` element that has a value unique within the enclosing `<unit>` element (with the exception of the matching inline elements in the `<source>`)
no prefix	an id for a `<segment>` or an `<ignorable>` or an inline element in the `<source>` element that has a value unique within the enclosing `<unit>` element (with the exception of the matching inline elements in the `<target>`)

Fragment identification process

Fragment identifiers are declared from the outside in. That is, a fragment identifier will first identify the <file> element, followed by the <group> and <unit> elements, in that order. Fragment identifiers may also be used for the <note>, , <ignorable>, and <data> elements as well as source and target inline elements. Only one of these additional elements may be used in an identifier, and it must be the last selector in the expression.

 A selector is the combination of a prefix and its ID. For example, this selector #f=f1 refers to the element <file id="f1">.

Selectors for modules and extensions must use a registered prefix (that is, an XML namespace prefix defined in the XLIFF file), and the value of the ID must be unique within the immediate enclosing <file>, <group>, or <unit> element.

Fragment identifiers may refer to an absolute path or a relative path. An absolute path fragment identifier starts with a forward slash " / " and expresses the absolute position of the fragment in that file. The forward slash may be preceded by a filename, in which case the absolute path starts at the root element of that file. As with an absolute path for a file name, an absolute fragment identifier must include all selectors between the root and the ID.

Fragment identifiers that do not start with a forward slash " / " are relative fragment identifiers. The scope of a relative fragment identifier is limited to the nearest ancestor element among <file>, <group>, and <unit>, unless the identifier contains a selector that unambiguously points to another file, group, or unit.

Example 14.4 contains examples of fragment identifiers. Each id attribute in the example is accompanied by either a custom attribute (xmrk:FragID) or an XML comment that contains the absolute fragment identifier that matches that value of that id attribute.

In Example 14.4 the attribute/value pair ref="g1" on the last <source> element (line 41) resolves to the <glossEntry> element that contains the ID g1 and not to the <group> element with the same ID value. This is because the scope of this relative reference is limited to the ancestor <unit> element. However, if a relative fragment identifier includes a selector for a <file>, <group>, or <unit> element, it can refer to an ID outside of its ancestor elements. In Example 14.4, if the <source> element contained the attribute/value pair ref="#g=g1/", the reference would resolve to <group id="g1">.

Example 14.4 – Fragment identifiers

```
1  <file id="seg" xmrk:FragID="my_file.xlf/#f=seg">
2    <unit id="seg1" xmrk:FragID="my_file.xlf/#f=seg/u=seg1">
3      <notes>
4        <note id="note1"
5          xmrk:FragID="my_file.xlf/#f=seg/u=seg1/n=note1">This
6          is a sample XLIFF 2.0 file.</note>
7      </notes>
8      <segment>
9        <source>Mr. Rodgers is a TV show from the U.S. not the U.K.
10         <pc id="pc1">
11         <!--Name: pc, FragID: my_file.xlf/#f=seg/u=seg1/pc1-->
12         Is Dr. Who from there U.K.?</pc>Yes: see the collected
13         viewer-ship on each show.<ph id="ph2" />
14         <!--Name: ph, FragID: my_file.xlf/#f=seg/u=seg1/ph2-->
15       </source>
16     </segment>
17   </unit>
18   <group id="g1" xmrk:FragID="my_file.xlf/#f=seg/g=g1">
19     <unit id="seg2"
20         xmrk:FragID="my_file.xlf/#f=seg/g=g1/u=seg2">
21       <gls:glossary>
22         <gls:glossEntry id="g1"
23           xmrk:FragID="my_file.xlf/#f=seg/g=g1/u=seg2/gls=g1">
24           <gls:term source="Termbase">Pitch</gls:term>
25           <gls:translation id="1" source="corpTermbase"
26             xmrk:FragID="my_file.xlf/#f=seg/g=g1/u=seg2/gls=1">
27             Tonhöhe</gls:translation>
28           <gls:definition
29             source="https://www.google.com/search?q=define+pitch">
30             the quality of a sound governed by the rate of
31             vibrations producing it; the degree of highness
32             or lowness of a tone.</gls:definition>
33         </gls:glossEntry>
34       </gls:glossary>
35       <xmrk:bookmark id="xmbm01"
36         xmrk:FragID="my_file.xlf/#f=seg/g=g1/u=seg2/xmrk=xmbm01">
37         <term>Guitar</term>
38       </xmrk:bookmark>
39       <segment>
40         <source>The guitar has great
41         <mrk id="m1" ref="#g1" type="term">
42         <!--Name: mrk, FragID: my_file.xlf/#f=seg/g=g1/u=seg2/m1-->
43         Pitch</mrk>.</source>
44       </segment>
45     </unit>
46   </group>
47 </file>
```

The XLIFF file in Example 14.5 contains two duplicate IDs. These are legal according to the XLIFF specification. However, to reference either one of these IDs, you must use the fragment identifier syntax to avoid ambiguity.

The reference `ref="my_file.xlf/#f=f1/u=1/gls=g1"` on line 17 unambiguously refers to the element `<gls:glossEntry id="g1">` (line 7) and not `<unit id="g1">` (line 25).

Example 14.5 – Fragment identifier example with duplicate IDs

```
1  <xliff xmlns="urn:oasis:names:tc:xliff:document:2.0" version="2.0"
2       srcLang="en" trgLang="pt"
3       xmlns:gls="urn:oasis:names:tc:xliff:glossary:2.0">
4    <file id="f1">
5      <unit id="1">
6        <gls:glossary>
7          <gls:glossEntry id="g1" ref="#m1">
8            <gls:term>bill</gls:term>
9            <gls:translation>Schnabel</gls:translation>
10           <gls:definition source="dictionary.com">The hard usually pointed
11               parts that cover a bird's mouth.</gls:definition>
12         </gls:glossEntry>
13       </gls:glossary>
14       <segment>
15         <source>It was a unique
16           <mrk id="m1" type="term"
17               ref="my_file.xlf/#f=f1/u=1/gls=g1">bill.</mrk>.
18         </source>
19         <!-- note, the ref attribute above shows the absolute fragment
20              identification. In many cases (including this example),
21              a reference to just the term will suffice, such as
22              ref="gls=g1" -->
23       </segment>
24     </unit>
25     <unit id="g1">
26       <segment>
27         <source>Many birds have unique bills.</source>
28       </segment>
29     </unit>
30   </file>
31 </xliff>
```

CHAPTER 15

Extensibility

XLIFF 2.0 supports limited extensibility. Extensibility must be used only to perform tasks not otherwise supported in the specification. This restriction is enforced in the processing requirements. We have used extensibility in this book to build a skeleton (see Chapter 9, the section titled "The minimalist method") and to add comments (see Chapter 14, *Fragment Identification*).

 Using extensibility can introduce risks to interoperability. An extension implemented by one XLIFF tool may not work seamlessly with another tool, even if the extension supports a feature not covered in the XLIFF specification.

Extensibility

You can extend XLIFF 2.0 in three ways:

- You can add elements from a custom namespace
- You can add attributes from a custom namespace
- You can use the metadata element module (see Chapter 19)

Custom elements and custom attributes

Custom elements and attributes may only be used inside the following elements:

- `<xliff>`
- `<file>`
- `<group>`
- `<unit>`
- `<note>`
- `<mrk>`
- `<sm>`

Consider a case where you need to track the version of a translation, perhaps for legal reasons, in a repository. The XLIFF 2.0 standard does not include a feature that supports interaction with repositories, so you can use extensibility.

To add a custom element or attribute, you first add the custom XML namespace, which can be any namespace that is not an XLIFF namespace. For this example, we will use the xmrk namespace that we have used previously:

Example 15.1 – Custom namespace declaration

```
<xliff xmlns="urn:oasis:names:tc:xliff:document:2.0"
       xmlns:xmrk="urn:xmarker"
```

Then, you add the custom element or attribute, which in Example 15.2 is the element `<xmrk:archive_note>` on line 5.

Example 15.2 – Custom element

```
 1 <?xml version="1.0" encoding="UTF-8"?>
 2 <xliff xmlns="urn:oasis:names:tc:xliff:document:2.0"
 3        xmlns:xmrk="urn:xmarker" srcLang="en" version="2.0">
 4   <file original="xliff_example.html" id="f1">
 5     <xmrk:archive_note>Initial version</xmrk:archive_note>
 6     <unit id="1">
 7       <segment>
 8         <source>History</source>
 9       </segment>
10     </unit>
11   </file>
12 </xliff>
```

The XLIFF 2.0 specification states that an extension must not provide the same functionality as an existing XLIFF core or module feature. However, you may extend or complement an existing feature. For example, you might augment the glossary module by adding a flag that shows the level of legal review required for an entry. The glossary module defined in XLIFF 2.0 does not support such a flag, so adding a custom attribute for this function would be compliant with the standard. In Example 15.3, we add the attribute xmrk:legal_approval_req="lax" to the <gls:glossEntry> (line 9).

Example 15.3 – Custom attribute

```
1  <?xml version="1.0"?>
2  <xliff xmlns="urn:oasis:names:tc:xliff:document:2.0"
3         version="2.0" srcLang="en" trgLang="pt"
4         xmlns:gls="urn:oasis:names:tc:xliff:glossary:2.0"
5         xmlns:xmrk="urn:xmarker">
6   <file id="f1">
7    <unit id="1">
8     <gls:glossary>
9      <gls:glossEntry id="g1" ref="#m1" xmrk:legal_approval_req="lax">
10      <gls:term>bill</gls:term>
11      <gls:translation>Schnabel</gls:translation>
12      <gls:definition source="dictionary.com">The hard usually
13          pointed parts that cover a bird's mouth.</gls:definition>
14     </gls:glossEntry>
15    </gls:glossary>
16    <segment>
17     <source>It was a unique
18       <mrk id="m1" type="term" ref="#g1">bill.</mrk>.
19     </source>
20    </segment>
21   </unit>
22   <unit id="g1">
23    <segment>
24     <source>Many birds have unique bills.</source>
25    </segment>
26   </unit>
27  </file>
28 </xliff>
```

Processors that do not support an extension should preserve the elements and attributes for that extension. However, the XLIFF 2.0 specification does not prohibit a processor from removing or modifying an extension.

Proposing extensions to XLIFF

XLIFF's extensibility provides a means for implementing new core or module features. The XLIFF Technical Committee[1] welcomes proposals for extensions and encourages interested people to participate in the work of the committee. If you have extended XLIFF and have a compliant XML schema, we encourage you to join the committee and bring your proposal.

_____

[1] http://www.oasis-open.org/committees/xliff/

XLIFF Modules

Beyond the XLIFF 2.0 Core, XLIFF defines a set of modules that provide optional features. Each module has its own namespace and XML schema. Tools and processes that support a module must comply with that module's constraints and processing instructions and must process XLIFF files according to the schemas and constraints defined by the XLIFF 2.0 specification.

These XLIFF modules define markup to support processing and define some constraints on supporting tools, however, they do not specify the manner in which these tools work or the output they must generate.

CHAPTER 16
Translation Candidates Module

The Translation Candidates module specifies a way to embed translation candidates generated from translation memory or a machine translation application in an XLIFF file and assign metadata to those candidates that captures match quality, match suitability, and similarity.

This module uses the elements `<matches>`, `<match>`, `<mda:metadata>`, `<xlf:originalData>`, `<xlf:source>`, and `<xlf:target>`; and the attributes id, matchQuality, matchSuitability, origin, ref, reference, similarity, subType, and type.

The Translation Candidates module uses the prefix `mtc` and the namespace `urn:oasis:names:tc:xliff:matches:2.0`.

Consider Example 16.1, Example 16.2, and Example 16.3, which outline a translation candidates workflow.

Example 16.2 shows a TMX file[1] that contains previously translated content related to the topic of Example 16.1.

For each segment in the TMX file that is deemed to be a 100% match against a segment in the XLIFF file, we provide a translation candidate match. Example 16.3 shows the resulting XLIFF file with translation candidates included.

With the match completed and the matches embedded, a translator or an automated process could score the matches for match quality, suitability, and similarity and assign a score, typically a decimal value between 0.0 and 100.0.

This process could be repeated with other translation resources such as a machine translation.

---

[1] Translation Memory eXchange (TMX) is a standard originally developed by LISA (Localization Industry Standards Association) and now maintained by GALA (Globalization and Localization Association). You can find the latest version at: http://www.gala-global.org/oscarStandards/tmx/tmx14b.html. The TMX standard is used to represent and exchange the contents of translation memory.

Example 16.1 – XLIFF file for translation candidates workflow

```
<xliff xmlns="urn:oasis:names:tc:xliff:document:2.0"
       version="2.0" srcLang="en" xmlns:xmrk="urn:xmarker"
       xmlns:mtc="urn:oasis:names:tc:xliff:matches:2.0">
  <file id="my1">
    <unit id="u1">
      <notes>
        <note id="note1">Hello.</note>
      </notes>
      <segment id="seg1">
        <source>Nirvana was an American grunge band from Aberdeen in the
        state of Washington.</source>
      </segment>
    </unit>
    <unit id="u2">
      <segment>
        <source>The Ramones were an American punk rock band that formed
        in the New York City neighborhood of Forest Hills, Queens, in
        1974.</source>
      </segment>
      <segment>
        <source>The
        <mrk id="m1" type="term">Clash</mrk>, founded in London in 1976,
            is considered one of the most influential early punk bands
            alongside other bands such as the Ramones and the Sex
            Pistols.</source>
      </segment>
    </unit>
    <unit id="u3">
      <segment>
        <source>Soundgarden is an American grunge - band from
        Seattle.</source>
      </segment>
    </unit>
    <unit id="u4">
      <segment>
        <source>The MC5 kicked off the whole
          <mrk id="m2" type="term">proto-punk</mrk>scene in Detroit.
        </source>
      </segment>
      <segment>
        <source>The MC5 album "Kick Out The Jams" was recorded live at a
        <mrk id="m3" type="term">concert</mrk> at Detroit's Grande
        Ballroom, October 1969.</source>
      </segment>
    </unit>
  </file>
</xliff>
```

Example 16.2 – Sample TMX file

```
<tmx version="1.4b">
  <header creationtool="MyTool" creationtoolversion="0.04"
        datatype="PlainText" segtype="sentence"
        adminlang="en-us" srclang="en" o-tmf="Some"></header>
  <body>
    <tu>
      <tuv xml:lang="en">
        <seg>The Clash, founded in London in 1976, is considered one of
        the most influential early punk bands alongside other bands such
        as the Ramones and the Sex Pistols.</seg>
      </tuv>
      <tuv xml:lang="de">
        <seg>The Clash (engl. „Der Zusammenprall"), gegründet 1976
        in London, gilt als eine der einflussreichsten frühen Punkbands
        neben anderen Bands wie den Ramones und den Sex Pistols.</seg>
      </tuv>
    </tu>
    <tu>
      <tuv xml:lang="en">
        <seg>Nirvana was an American grunge band from Aberdeen in the
        state of Washington.</seg>
      </tuv>
      <tuv xml:lang="de">
        <seg>Nirvana war eine US-amerikanische Grunge-Band aus Aberdeen
        im Bundesstaat Washington.</seg>
      </tuv>
    </tu>
    <tu>
      <tuv xml:lang="en">
        <seg>Soundgarden is an American grunge - band from Seattle.</seg>
      </tuv>
      <tuv xml:lang="de">
        <seg>Soundgarden ist eine US-amerikanische Grunge-Band aus
        Seattle.</seg>
      </tuv>
    </tu>
  </body>
</tmx>
```

Example 16.3 – XLIFF file with translation candidates included

```
<xliff xmlns="urn:oasis:names:tc:xliff:document:2.0"
       version="2.0" srcLang="en"
       trgLang="de" xmlns:xmrk="urn:xmarker"
       xmlns:mtc="urn:oasis:names:tc:xliff:matches:2.0">
  <file id="my1">
    <unit xmlns:xlf="urn:oasis:names:tc:xliff:document:2.0" id="u1">
      <mtc:matches>
        <mtc:match ref="#m2-">
          <source>Nirvana was an American grunge band from Aberdeen in
            the state of Washington.</source>
          <target>Nirvana war eine US-amerikanische Grunge-Band aus
            Aberdeen im Bundesstaat Washington.</target>
        </mtc:match>
      </mtc:matches>
      <notes>
        <note id="note1">Hello.</note>
      </notes>
      <segment id="seg1">
        <source>
          <mrk id="m2-" type="mtc:match">Nirvana was an American
          grunge band from Aberdeen in the state of Washington.</mrk>
        </source>
      </segment>
    </unit>
    <unit xmlns:xlf="urn:oasis:names:tc:xliff:document:2.0" id="u2">
      <mtc:matches>
        <mtc:match ref="#m2-">
          <source>The Clash founded in London in 1976, is considered
            one of the most influential early punk bands alongside
            other bands such as the Ramones and the Sex Pistols.
          </source>
          <target>The Clash (engl. „Der Zusammenprall"), gegründet
            1976 in London, gilt als eine der einflussreichsten
            frühen Punkbands neben anderen Bands wie den Ramones und
            den Sex Pistols.
          </target>
        </mtc:match>
      </mtc:matches>
      <segment>
        <source>The Ramones were an American punk rock band that formed
          in the New York City neighborhood of Forest Hills, Queens, in
          1974.</source>
      </segment>
      <segment>
        <source>
          <mrk id="m2-" type="mtc:match">The
          <mrk id="m1" type="term">Clash</mrk>founded in London in 1976,
            is considered one of the most influential early punk bands
```

```
        alongside other bands such as the Ramones and the Sex
        Pistols.</mrk>
      </source>
  </unit>
  <unit xmlns:xlf="urn:oasis:names:tc:xliff:document:2.0" id="u3">
    <mtc:matches>
      <mtc:match ref="#m1-">
        <source>Soundgarden is an American grunge - band from
          Seattle.</source>
        <target>Soundgarden ist eine US-amerikanische Grunge-Band
          aus Seattle.</target>
      </mtc:match>
    </mtc:matches>
    <segment>
      <source>
        <mrk id="m1-" type="mtc:match">Soundgarden is an American
          grunge - band from Seattle.</mrk>
      </source>
    </segment>
  </unit>
  <unit xmlns:xlf="urn:oasis:names:tc:xliff:document:2.0" id="u4">
    <segment>
      <source>The MC5 kicked off the whole
        <mrk id="m2" type="term">proto-punk</mrk>scene in Detroit.
      </source>
    </segment>
    <segment>
      <source>The MC5 album "Kick Out The Jams" was recorded
        live at a <mrk id="m3" type="term">concert</mrk>at
        Detroit's Grande Ballroom, October 1969.</source>
    </segment>
  </unit>
  </file>
</xliff>
```

CHAPTER 17
Glossary Module

The Glossary module lets you embed glossary entries an XLIFF file. A glossary can be helpful for managing terms, keywords, and product names and helping to ensure consistent use of terminology across an organization's content. This module defines methods for adding definitions and translations for terms using the elements `<glossary>`, and `<definition>` and the attributes id, ref, and source.

The Glossary module uses the namespace urn:oasis:names:tc:xliff:glossary:2.0 and the prefix gls.

This chapter carries on where Chapter 16, *Translation Candidates Module*, left off. In Chapter 16, potential matches from a TMX file were embedded in an XLIFF file using the Translation Candidates module. For glossary terms, we will process that XLIFF file (Example 16.3, "XLIFF file with translation candidates included") using a TBX file.[1]

The preprocessing step compares the terms in Example 16.3 with terms in a TBX file and embeds annotations for each matching term in the XLIFF file using the markup defined by the Glossary module. Example 17.1 shows a TBX file containing the terms "Clash" and "concert."

Example 17.2 shows the merged XLIFF file. In this example, the processor finds each term (`<mrk type="term">`) in the XLIFF file, searches for that term in the TBX file, and if it finds a match, it inserts the glossary information from the TBX file into the parent `<unit>` element in the XLIFF file and generates a reference from the `<gls:glossEntry>` element to the instance of the term.

---

[1] TermBase eXchange (TBX) is an ISO standard (ISO 30042:2008) for representing terminology-related information.

Example 17.1 – Glossary workflow: TBX file

```xml
<martif type="TBX" xml:lang="en-us">
  <martifHeader>
    <fileDesc>
      <sourceDesc><p>Music terms</p></sourceDesc>
    </fileDesc>
      <encodingDesc>
        <p type="DCSName"> SYSTEM "myTBX.xml" </p>
      </encodingDesc>
  </martifHeader>
  <body>
    <termEntry id="t1">
      <descrip type="subjectField">Band</descrip>
      <descrip type="conceptPosition" target="foodskos">s79</descrip>
      <descrip type="definition">Early punk band from England.</descrip>
      <langSet xml:lang="en-us">
        <ntig id="n1">
          <termGrp><term>Clash</term></termGrp>
        </ntig>
      </langSet>
      <langSet xml:lang="de">
        <ntig id="n2">
          <termGrp><term>Clash</term></termGrp>
        </ntig>
      </langSet>
    </termEntry>
    <termEntry id="t2">
      <descrip type="subjectField">Music Event</descrip>
      <descrip type="conceptPosition" target="foodskos">s81</descrip>
      <descrip type="definition">a live performance (typically of music)
        before an audience</descrip>
      <langSet xml:lang="en-us">
        <ntig id="n3">
          <termGrp><term>concert</term></termGrp>
        </ntig>
      </langSet>
      <langSet xml:lang="de">
        <ntig id="n4">
          <termGrp><term>Konzert (Veranstaltung)</term></termGrp>
        </ntig>
      </langSet>
    </termEntry>
  </body>
</martif>
```

Example 17.2 – Glossary workflow: merged file

```
<xliff xmlns="urn:oasis:names:tc:xliff:document:2.0"
       version="2.0" srcLang="en"
       trgLang="de" xmlns:xmrk="urn:xmarker"
       xmlns:gls="urn:oasis:names:tc:xliff:glossary:2.0"
       xmlns:mtc="urn:oasis:names:tc:xliff:matches:2.0">
<file id="my1">
  <unit xmlns:xlf="urn:oasis:names:tc:xliff:document:2.0" id="u1">
    <mtc:matches>
      <mtc:match ref="#m2-">
        <source>Nirvana was an American grunge band from Aberdeen in the
          state of Washington.</source>
        <target>Nirvana war eine US-amerikanische Grunge-Band aus
          Aberdeen im Bundesstaat Washington.</target>
      </mtc:match>
    </mtc:matches>
    <notes><note id="note1">Hello.</note></notes>
    <segment id="seg1">
      <source>
        <mrk id="m2-" type="mtc:match">Nirvana was an American grunge
          band from Aberdeen in the state of Washington.</mrk>
      </source>
    </segment>
  </unit>
  <unit xmlns:xlf="urn:oasis:names:tc:xliff:document:2.0" id="u2">
    <gls:glossary>
      <gls:glossEntry ref="#m1">
        <gls:term source="publicTermbase">Clash</gls:term>
        <gls:translation id="1" source="myTermbase">Clash
        </gls:translation>
        <gls:definition source="publicTermbase">Early punk band
          from England.</gls:definition>
      </gls:glossEntry>
    </gls:glossary>
    <mtc:matches>
      <mtc:match ref="#m2-">
        <source>The Clash founded in London in 1976, is considered one
          of the most influential early punk bands alongside other bands
          such as the Ramones and the Sex Pistols.</source>
        <target>The Clash (engl. „Der Zusammenprall"), gegründet
          1976 in London, gilt als eine der einflussreichsten frühen
          Punkbands neben anderen Bands wie den Ramones und den Sex
          Pistols.</target>
      </mtc:match>
    </mtc:matches>
    <segment>
      <source>The Ramones were an American punk rock band that formed
        in the New York City neighborhood of Forest Hills, Queens,
        in 1974.</source>
```

```
        <segment>
          <source>
            <mrk id="m2-" type="mtc:match">The
            <mrk id="m1" type="term">Clash</mrk>, founded in London in 1976,
              is considered one of the most influential early punk bands
              alongside other bands such as the Ramones and the Sex Pistols.
            </mrk>
          </source>
        </segment>
      </unit>
      <unit xmlns:xlf="urn:oasis:names:tc:xliff:document:2.0" id="u3">
        <mtc:matches>
          <mtc:match ref="#m1-">
            <source>Soundgarden is an American grunge - band from
              Seattle.</source>
            <target>Soundgarden ist eine US-amerikanische Grunge-Band
              aus Seattle.</target>
          </mtc:match>
        </mtc:matches>
        <segment>
          <source>
            <mrk id="m1-" type="mtc:match">Soundgarden is an American
              grunge - band from Seattle.</mrk>
          </source>
        </segment>
      </unit>
      <unit xmlns:xlf="urn:oasis:names:tc:xliff:document:2.0" id="u4">
        <gls:glossary>
          <gls:glossEntry ref="#m3">
            <gls:term source="publicTermbase">concert</gls:term>
            <gls:translation id="1" source="myTermbase">Konzert
              (Veranstaltung)</gls:translation>
            <gls:definition source="publicTermbase">a live performance
              (typically of music) before an audience</gls:definition>
          </gls:glossEntry>
        </gls:glossary>
        <segment>
          <source>The MC5 kicked off the whole
            <mrk id="m2" type="term">proto-punk</mrk>scene in Detroit.
          </source>
        </segment>
        <segment>
          <source>The MC5 album "Kick Out The Jams" was recorded
            live at a <mrk id="m3" type="term">concert</mrk> at
            Detroit's Grande Ballroom, October 1969.</source>
        </segment>
      </unit>
    </file>
</xliff>
```

CHAPTER 18
Format Style Module

The Format Style module defines two attributes, `fs` and `subFs`, that let you associate an HTML element and optional attributes with an XLIFF element. Translation support software can then use the HTML element to render a more friendly, browser view of the XLIFF file for translators and reviewers. The Format Style module also provides a standard method for capturing this information to help ensure interoperability between tools.

The module specification defines a list of HTML elements that can be used in this context. The list includes a wide variety of formatting and simple structural HTML elements but does not allow processing elements, such as `<script>`.

The `fs` attribute holds the name of an HTML element. The optional `subFs` attribute holds one or more comma-separated attribute/value pairs that will be applied to the HTML element. You can use this for required attributes, such as the `src` attribute on the `<img>` element.

 It is a violation of the XLIFF 2.0 processing requirements to use the Format Style module to perform a round-trip merge.

The Format Style module uses the namespace `urn:oasis:names:tc:xliff:fs:2.0` and the prefix `fs`.

Example 18.1 – Format style workflow: original XLIFF file

```
<file id="f1">
  <unit id="1">
   <segment>
    <source>Mick Jones renewed his interest in the Vintage
        <pc id="1">'72 Telecaster Thinline </pc> guitar.
        <ph id="ph2" />He says <pc id="2">I love 'em</pc>
        <ph id="ph1" /></source>
   </segment>
   </unit>
</file>
```

Example 18.1 shows the original XLIFF file.

In Example 18.2, we put `fs:fs` and `fs:subFs` attributes on the elements that will be rendered in the browser.

Example 18.2 – Format style workflow: XLIFF file with `fs` and `subFs` attributes

```
<file id="f1" fs:fs="html">
 <unit id="1" fs:fs="p">
  <segment>
   <source>Mick Jones renewed his interest in the Vintage
    <pc id="1" fs:fs="strong">'72 Telecaster Thinline </pc> guitar.
    <ph id="ph2" fs:fs="br" />He says <pc id="2" fs:fs="q">I love 'em</pc>
    <ph id="ph1" fs:fs="img" fs:subFs="src,smileface.png" /></source>
  </segment>
 </unit>
</file>
```

You can use a script or an XSL transform to convert the XLIFF into HTML (see Example 18.3) for display. The generated HTML must be a valid HTML snippet. Any XLIFF element that does not contain an `fs:fs` attribute should be removed in the conversion.

Example 18.3 – Format style workflow: generated HTML file

```
<html>
  <p>Mick Jones renewed his interest in the Vintage
    <strong>'72 Telecaster Thinline </strong> guitar.
    <br/>He says <q>I love 'em</q> <img src="smileface.png"/>
  </p>
</html>
```

CHAPTER 19

Metadata Module

The Metadata module lets you add custom metadata to XLIFF 2.0 files without using a custom namespace. It does this with the elements `<metadata>`, `<metaGroup>`, and `<meta>`; and the attributes `id`, `category`, `appliesTo`, and `type`.

The Metadata module uses the namespace `urn:oasis:names:tc:xliff:metadata:2.0` and the prefix `mda`.

Consider the following case. You need to translate an automated email message that uses the Sendmail smtp server syntax and create a ready-to-send message. Normally, you would need two passes to do this: one pass by a merger agent to compile the translation and another pass from an agent to create the message. By embedding the Sendmail metadata in the XLIFF file, you can translate and create the message in a single-pass. Example 19.1 shows the initial XLIFF file.

Example 19.1 – Initial XLIFF file for metadata module example

```
<file id="my1">
  <unit id="1">
    <segment>
      <source>Java is an object oriented programming language and a
        registered trademark of Sun Microsystems
        (bought by Oracle in 2010).
      </source>
    </segment>
  </unit>
</file>
```

Example 19.2 shows the same file after translation and after the Sendmail metadata has been added.

Example 19.2 – Example 19.1 with Sendmail metadata added

```
<file id="my1">
  <unit id="1">
    <mda:metadata id="m1">
      <mda:metaGroup id="mg1" category="sendmail metadata"
                     appliesTo="target">
      <mda:meta type="From">Translation-coordinator@xmarker.com</mda:meta>
      <mda:meta type="To">java-department@xmarker.com</mda:meta>
      <mda:meta type="MIME-Version">1.0</mda:meta>
      <mda:meta type="Content-Type">multipart/alternative;</mda:meta>
      <mda:meta type="boundary">
         'PAA07345.1079345986/server.xmarker.com'</mda:meta>
      <mda:meta type="Subject">Translation approval</mda:meta>
      <mda:meta type="header message">
         This is a MIME-encapsulated message</mda:meta>
      <mda:meta type="upper delim">
          --PAA07345.1079345986/server.xmarker.com</mda:meta>
      <mda:meta type="Content-Type">text/html</mda:meta>
      <mda:meta type="lower delim">
        --PAA07345.1079345986/server.xmarker.com</mda:meta>
      <mda:meta type="command"> | sendmail -t</mda:meta>
      </mda:metaGroup>
    </mda:metadata>
    <segment>
      <source>Java is an object oriented programming language and a
      registered trademark of Sun Microsystems
      (bought by Oracle in 2010).
      </source>
      <target>Java ist eine objektorientierte Programmiersprache und
      eine eingetragene Marke des Unternehmens Sun Microsystems
      (2010 von Oracle aufgekauft).
      </target>
    </segment>
  </unit>
</file>
```

Example 19.2 can then be processed to create a ready-to-send email message. The software that processes the file must understand how to create a valid Sendmail smtp message, but the XLIFF file contains all of the metadata needed to do that. Example 19.3 shows the ready-to-send message as a command line that can be directly executed to send the message.

Example 19.3 – Mail message created from the XLIF file in Example 19.2

```
echo " From: Translation-coordinator@xmarker.com
To: java-department@xmarker.com
MIME-Version: 1.0
Content-Type: multipart/alternative;
boundary='PAA07345.1079345986/server.xmarker.com'
Subject: Translation approval

This is a MIME-encapsulated message

--PAA07345.1079345986/server.xmarker.com
Content-Type: multipart/alternative;

<html>
<head>
  <title>Please approve translation</title>
</head>
<body>
  <p>Source: Java is an object oriented programming language
    and a registered trademark of the company Sun Microsystems
    (bought by Oracle in 2010).</p>
  <p>Target: Java ist eine objektorientierte Programmiersprache
    und eine eingetragene Marke des Unternehmens Sun Microsystems
    (2010 von Oracle aufgekauft).</p>
</body>
</html>
--PAA07345.1079345986/server.xmarker.com
" | sendmail -t
```

CHAPTER 20
Resource Data Module

The Resource Data module provides a mechanism for referencing external resource data that may need to be modified or used as a contextual reference during translation. This module defines the elements `<resourceData>`, and `<reference>` and the attributes `id`, `xml:lang`, `mimeType`, `context`, `href`, and `ref`.

The Resource Data module uses the prefix `res` and the namespace `urn:oasis:names:tc:xliff:resourcedata:2.0`.

Consider the following case: firmware for an instrument stores error codes and messages in a config file. You need a German translation of the config file, so you extract the text into an XLIFF 2.0 file using an extractor agent.

A traditional method might have three steps: 1) Translate the XLIFF. 2) Merge the file back to its original format. 3) Post-process the merged file into the format required for the config file.

You can remove one step by using the Resource Data module. By embedding the target config file's resource data, you can directly compile the translated XLIFF file into the instrument's firmware.

We begin with the English language config file (see Example 20.1):

Example 20.1 – English language config file

```xml
<?xml version="1.0" encoding="utf-8"?>
<resource category="error strings">
 <string id="s1">class library, interface, or enum expected</string>
 <string id="s2">illegal start of expression</string>
</resource>
```

We then extract the source strings into an XLIFF file and, using the Resource Data module markup, include the information needed to compile the firmware. We then translate the XLIFF, which gives us Example 20.2.

Example 20.2 – XLIFF file with Resource Data module markup

```
 1  <file id="f1">
 2      <res:resourceData>
 3        <res:resourceItem id="r1" mimeType="text/xml" context="no">
 4          <res:source href="../resource_files/resource_en.xml" />
 5          <res:target href="../resource_files/resource_de.xml" />
 6        </res:resourceItem>
 7      </res:resourceData>
 8      <unit id="u1" name="s1">
 9        <res:resourceData>
10          <res:resourceItemRef ref="r1" />
11        </res:resourceData>
12        <segment id="1" state="translated">
13          <source>class library, interface, or enum expected</source>
14          <target>Klassenbibliothek, Schnittstellen, oder Aufzählung
15            erwartet</target>
16        </segment>
17      </unit>
18      <unit id="u2" name="s2">
19        <res:resourceData>
20          <res:resourceItemRef ref="r1" />
21        </res:resourceData>
22        <segment id="1" state="translated">
23          <source>illegal start of expression</source>
24          <target>illegale Beginn der Ausdruck</target>
25        </segment>
26      </unit>
27  </file>
```

In Example 20.2, the `<res:resourceData>` element (line 2) captures the names of the source and target config files in the `<res:resourceItem>` element (lines 3–6). On lines 9–11 and 19–21, the `<res:resourceData>` elements refer to the `<res:resourceItem>` element, and the name attribute on each `<unit>` element captures the id for that string in the source file. This gives a merger agent enough information to construct a valid target file with the translated content (see Example 20.3).

Example 20.3 – Translated config file

```
<?xml version="1.0" encoding="utf-8"?>
<resource>
  <string id="s1">Klassenbibliothek, Schnittstellen, oder Aufzählung
    erwartet</string>
  <string id="s2">illegale Beginn der Ausdruck</string>
</resource>
```

CHAPTER 21
Change Tracking Module

The Change Tracking module lets you store revision information in an XLIFF 2.0 file. This capability can help you track changes when you have more than one translator on a project or when your workflow includes translation memory or machine language steps.

This module uses the elements `<ctr:changeTrack>`, `<ctr:revisions>`, `<ctr:revision>`, `<ctr:item>` and the attributes `appliesTo`, `ref`, `currentVersion`, `author`, `datetime`, `version`, and `property`.

The Change Tracking module uses the prefix `ctr` and the namespace `urn:oasis:names:tc:xliff:changetracking:2.0`.

By comparing a current version of an XLIFF 2.0 file to past versions, a user can see what changes have been made in the lifecycle of the file and, potentially, see additional information such as who made the changes and when the changes were made.

Consider the following case. You are translating the second version of an XLIFF 2.0 file, and you want to track the changes that have taken place since the initial version. Example 21.1 shows version 2 of the XLIFF file.

Example 21.1 – Change Tracking module: sample file – version 2

```
<file id="my1">
  <unit id="u1">
    <notes><note id="note1">Hello again.</note></notes>
    <segment id="seg1">
      <source>
        <mrk id="m2-u1" type="mtc:match">Nirvana was an American
          grunge band from Aberdeen in the state of Washington.</mrk>
      </source>
      <target>Nirvana war eine US-amerikanische Grunge-Band aus
        Aberdeen im Bundesstaat Washington.</target>
    </segment>
  </unit>
  <unit id="u2">
    <segment>
      <source>The Ramones were an American punk rock band that
      formed in the New York City neighborhood of Forest Hills,
      Queens, in 1974.</source>
```

```
        <segment>
          <source>
            <mrk id="m2-u2" type="mtc:match">The
            <mrk id="m1" type="term">Clash</mrk>, founded in London in 1976,
                is considered one of the most influential early punk bands
                    alongside other bands such as the Ramones and the Sex Pistols.
            </mrk>
          </source>
          <target>The Clash, gegründet 1976 in London, gilt als eine der
              einflussreichsten frühen Punkbands neben anderen Bands wie den
              Ramones und den Sex Pistols.</target>
        </segment>
      </unit>
      <unit id="u3">
        <segment>
          <source>
            <mrk id="m1-u3" type="mtc:match">Soundgarden is an American
                grunge - band from Seattle.</mrk>
          </source>
          <target>Soundgarden ist eine US-amerikanische Grunge-Band aus
              Seattle.</target>
        </segment>
      </unit>
      <unit id="u4">
        <segment>
          <source>The MC5 kicked off the whole
            <mrk id="m2" type="term">proto-punk</mrk>scene in Detroit,
              Michigan.</source>
        </segment>
        <segment>
          <source>The MC5 album "Kick Out The Jams" was recorded live at a
          <mrk id="m3" type="term">concert</mrk>at Detroit's Grande
              Ballroom, October 1969.</source>
        </segment>
      </unit>
    </file>
```

Example 21.2 shows version 1 of the file, which will be compared with version 2.

Example 21.2 – Change Tracking module: sample file – version 1

```
<file id="my1">
  <unit id="u1">
    <notes>
      <note id="note1">Hello.</note>
    </notes>
    <segment id="seg1">
```

```xml
    <source>
      <mrk id="m2-u1" type="mtc:match">Nirvana was an American
        grunge band from Aberdeen in the state of Washington.</mrk>
    </source>
    <target>Nirvana war eine US-amerikanische Grunge-Band aus
      Aberdeen im Bundesstaat Washington.</target>
</unit>
<unit id="u2">
  <segment>
    <source>The Ramones were an American punk rock band that
    formed in the New York City neighborhood of Forest Hills,
    Queens, in 1974.</source>
  </segment>
  <segment>
    <source>
      <mrk id="m2-u2" type="mtc:match">The
      <mrk id="m1" type="term">Clash</mrk>, founded in London in 1976,
        is considered one of the most influential early punk bands
        alongside other bands such as the Ramones and the Sex Pistols.
      </mrk>
    </source>
    <target>The Clash (engl. „Der Zusammenprall"), gegründet 1976
      in London, gilt als eine der einflussreichsten frühen Punkbands
      neben anderen Bands wie den Ramones und den Sex Pistols.
    </target>
  </segment>
</unit>
<unit id="u3">
  <segment>
    <source>
      <mrk id="m1-u3" type="mtc:match">Soundgarden is an American
        grunge - band from Seattle.</mrk>
    </source>
    <target>Soundgarden ist eine US-amerikanische Grunge-Band aus
      Seattle.</target>
  </segment>
</unit>
<unit id="u4">
  <segment>
    <source>The MC5 kicked off the whole proto-punk scene in
      Detroit.</source>
  </segment>
  <segment>
    <source>The MC5 album "Kick Out The Jams" was recorded live at a
    <mrk id="m3" type="term">concert</mrk>at Detroit's Grande
      Ballroom, October 1969.</source>
  </segment>
</unit>
</file>
```

Example 21.3 shows the result of comparing version 2 against version 1 and capturing the revisions using markup from the Change Tracking module.

Example 21.3 – Change Tracking module: marked file

```
<file id="my1">
  <unit id="u1">
    <ctr:changeTrack>
      <ctr:revisions appliesTo="note" ref="note1">
        <ctr:revision datetime="2013-07-15T10:00:00+8:00" author="Joe">
          <ctr:item property="content">Hello.</ctr:item>
        </ctr:revision>
      </ctr:revisions>
    </ctr:changeTrack>
    <notes><note id="note1">Hello again.</note></notes>
    <segment id="seg1">
      <source>
        <mrk id="m2-u1" type="mtc:match">Nirvana was an American
          grunge band from Aberdeen in the state of Washington.</mrk>
      </source>
      <target>Nirvana war eine US-amerikanische Grunge-Band aus
        Aberdeen im Bundesstaat Washington.</target>
    </segment>
  </unit>
  <unit id="u2">
    <ctr:changeTrack>
      <ctr:revisions appliesTo="target">
        <ctr:revision>
          <ctr:item property="content">The Clash (engl. „Der
            Zusammenprall"), gegründet 1976 in London, gilt als eine der
            einflussreichsten frühen Punkbands neben anderen Bands wie den
            Ramones und den Sex Pistols.</ctr:item>
        </ctr:revision>
      </ctr:revisions>
    </ctr:changeTrack>
    <segment>
      <source>The Ramones were an American punk rock band that
        formed in the New York City neighborhood of Forest Hills,
        Queens, in 1974.</source>
    </segment>
    <segment>
      <source>
        <mrk id="m2-u2" type="mtc:match">The
        <mrk id="m1" type="term">Clash</mrk>founded in London in 1976,
          is considered one of the most influential early punk bands
          alongside other bands such as the Ramones and the Sex Pistols.
        </mrk>
      </source>
      <target>The Clash, gegründet 1976 in London, gilt als eine der
```

```
        einflussreichsten frühen Punkbands neben anderen Bands wie den
        Ramones und den Sex Pistols.</target>
  </unit>
  <unit id="u3">
    <segment>
      <source>
        <mrk id="m1-u3" type="mtc:match">Soundgarden is an American
          grunge - band from Seattle.</mrk></source>
      <target>Soundgarden ist eine US-amerikanische Grunge-Band aus
        Seattle.</target>
    </segment>
  </unit>
  <unit id="u4">
    <ctr:changeTrack>
      <ctr:revisions appliesTo="source">
        <ctr:revision>
          <ctr:item property="content">The MC5 kicked off the whole
            proto-punk scene in Detroit.</ctr:item>
        </ctr:revision>
      </ctr:revisions>
    </ctr:changeTrack>
    <segment>
      <source>The MC5 kicked off the whole <mrk id="m2"
        type="term">proto-punk</mrk> scene in Detroit, Michigan.</source>
    </segment>
    <segment>
      <source>The MC5 album "Kick Out The Jams" was recorded live at a
        <mrk id="m3" type="term">concert</mrk>at Detroit's Grande
        Ballroom, October 1969.</source>
    </segment>
  </unit>
</file>
```

The following changes were made between version 1 and version 2.

- The contents of the first note in the first unit (id="u1") changed from "Hello." in version 1 to "Hello again." in version 2.
- The translation of the second segment in the second unit (id="u2") removed the string "(engl. „Der Zusammenprall")" from version 1.
- The first source line in the fourth unit (id="u4") said "The MC5 kicked off the whole proto-punk scene in Detroit." in version 1 and changed to "The MC5 kicked off the whole <mrk id="m2" type="term">proto-punk</mrk> scene in Detroit, Michigan." in version 2.

CHAPTER 22
Size and Length Restriction Module

The Size and Length Restriction module enables extractor and enricher agents to set size and length limits for text. You can use this capability to inform translators about space limitations for user interface text blocks and flag text that exceeds those limits. You can set restrictions on both the amount of space allowed to store text and the amount of space available to display text.

In this module, the term *storage* refers to the amount of space required to store text in memory, and the term *general* refers to an approximation of display width, which takes into account single- and multi-byte character widths but not factors such as font, point size, or ligatures.

This module uses the elements `<profiles>` and `<data>`; and the attributes `storageProfile`, `generalProfile`, `storage`, `general`, `profile`, `storageRestriction`, `sizeRestriction`, `equivStorage`, `sizeInfo` and `sizeInfoRef`.

The Size and Length Restriction module uses the prefix `slr` and the namespace `urn:oasis:names:tc:xliff:sizerestriction:2.0`.

This module defines a profile at either the `<file>` or `<unit>` level. The profile identifies the text encoding and any Unicode normalization. You define Unicode normalization using the `<normalization>` element, which can be set to nfc, nfd, or none.[1]

Consider a case where we want to restrict the length of a translation to fit the Twitter limit of 140 Unicode NFC code points. Begin with the XLIFF 2.0 file in Example 22.1, which contains English source text and a German translation. The Size and Length Restriction profile is set for UTF-8 NFC codepoints (lines 2–7), and each source and target has been given a size restriction of 140 characters (the `slr:sizeRestriction` attributes on lines 19, 25, 33, and 37).

Example 22.2 shows the result after the processor inserts comments (lines 25–26 and 39) that identify whether each target meets the restriction. The first target violates the restriction, but the second one does not. How a processor identifies problems is implementation specific. The specification simply defines the markup for specifying restrictions.

---

[1] The Unicode normalization forms are defined in the Unicode Standard Annex #15 http://unicode.org/reports/tr15/.

Example 22.1 – Size and Length Restriction module: original XLIFF file

```
 1 <file id="my1">
 2   <slr:profiles generalProfile="xliff:codepoints"
 3                   storageProfile="xliff:utf8">
 4     <!-- Select standard UTF-8 storage encoding and standard
 5         codepoint size restriction both with NFC normalization-->
 6     <slr:normalization general="nfc" storage="nfc" />
 7   </slr:profiles>
 8   <unit id="u1">
 9     <notes>
10      <note>This is a test to see if the translated strings will meeting
11         Twitter's 140 character maximum requirement.</note>
12      <note>For Unicode strings, Twitter assumes UTF-8 and uses
13         Normalization Form C (NFC).</note>
14      <note>This test will use NFC and report which translations will
15         tweet, and which will not.</note>
16     </notes>
17     <segment id="seg1">
18       <source>
19         <pc id="pc1" slr:sizeRestriction="140">The length of this
20           sentence should be acceptable in Twitter. This sentence
21           is not too long. This sentence is also not too long.
22           This is all.</pc>
23       </source>
24       <target>
25         <pc id="pc1" slr:sizeRestriction="140">Die Länge dieser Satz
26           sollte akzeptabel sein, in Twitter. Dieser Satz ist nicht
27           zu lang. Dieser Satz ist auch nicht zu lang. Das ist alles.
28           Das ist Deutsch.</pc>
29       </target>
30     </segment>
31     <segment id="seg2">
32       <source>
33         <pc id="pc2" slr:sizeRestriction="140">The complex of poetic
34           genres in the history.</pc>
35       </source>
36       <target>
37         <pc id="pc2" slr:sizeRestriction="140">Der Komplex der
38           poetischen Gattungen in der Geschichte.</pc>
39       </target>
40     </segment>
41   </unit>
42 </file>
```

Example 22.2 – Size and Length Restriction module: XLIFF file with result comments

```
1  <file id="my1">
2    <slr:profiles generalProfile="xliff:codepoints"
3                  storageProfile="xliff:utf8">
4      <!-- Select standard UTF-8 storage encoding and standard
5          codepoint size restriction both with NFC normalization-->
6      <slr:normalization general="nfc" storage="nfc" />
7    </slr:profiles>
8    <unit id="u1">
9      <notes>
10       <note>This is a test to see if the translated strings will meeting
11          Twitter's 140 character maximum requirement.</note>
12       <note>For Unicode strings, Twitter assumes UTF-8 and uses
13          Normalization Form C (NFC).</note>
14       <note>This test will use NFC and report which translations will
15          tweet, and which will not.</note>
16     </notes>
17     <segment id="seg1">
18       <source>
19         <pc id="pc1" slr:sizeRestriction="140">The length of this
20            sentence should be acceptable in Twitter. This sentence
21            is not too long. This sentence is also not too long.
22            This is all.</pc>
23       </source>
24       <target>
25         <!--size-length-restriction violation: maximum codepoint-length
26             allowed is 140 - but the codepoint-length is 156.-->
27         <pc id="pc1" slr:sizeRestriction="140">Die Länge dieser Satz
28            sollte akzeptabel sein, in Twitter. Dieser Satz ist nicht
29            zu lang. Dieser Satz ist auch nicht zu lang. Das ist alles.
30            Das ist Deutsch.</pc>
31       </target>
32     </segment>
33     <segment id="seg2">
34       <source>
35         <pc id="pc2" slr:sizeRestriction="140">The complex of poetic
36            genres in the history.</pc>
37       </source>
38       <target>
39         <!--no violation; the codepoint-length is 54.-->
40         <pc id="pc2" slr:sizeRestriction="140">Der Komplex der
41            poetischen Gattungen in der Geschichte.</pc>
42       </target>
43     </segment>
44   </unit>
45 </file>
```

CHAPTER 23
Validation Module

The Validation module provides a mechanism for including validation rules for target text in an XLIFF file. This mechanism is typically used for rules, such as style guidelines, that are not easy to validate with a parser. You can set rules globally or on individual segments, and you can make rules conditional based on the contents of the source. You can also override global rules for selected segments.

This module uses the elements `<validation>` and the attributes `isPresent`, `occurs`, `isNotPresent`, `startsWith`, `endsWith`, `existsInSource`, `caseSensitive`, `normalization`, and `disabled`.

The Validation module uses the namespace `urn:oasis:names:tc:xliff:validation:2.0` and the prefix `val`.

Example 23.1 contains 15 units, each of which contains the same source and target text but different validation rules.

Example 23.1 – Validation module: XLIFF file with validation rules

```
<file id="my1">
 <unit id="1">
   <segment id="1">
     <source>Choose an option in the online store:</source>
     <target>Escolha uma opção na loja online:</target>
   </segment>
 </unit>
 <unit id="2">
   <val:validation>
     <val:rule isPresent="store" disabled="yes" />
   </val:validation>
   <segment id="1">
     <source>Choose an option in the application store:</source>
     <target>Escolha uma opção na application store:</target>
   </segment>
 </unit>
 <unit id="v1">
  <val:validation>
   <val:rule isPresent="online" />
  </val:validation>
  <segment id="1">
```

```
    <source>Choose an option in the online store.</source>
    <target>Escolha uma opção na loja online.</target>
  </unit>
  <unit id="v2">
   <val:validation>
     <val:rule isPresent="loja" />
   </val:validation>
   <segment id="1">
     <source>Choose an option in the online store.</source>
     <target>Escolha uma opção na online store.</target>
   </segment>
  </unit>
  <unit id="v3">
   <val:validation>
     <val:rule isPresent="loja" occurs="2" />
   </val:validation>
   <segment id="1">
     <source>Choose a store option in the online store.</source>
     <target>Escolha uma opção de loja na loja online.</target>
   </segment>
  </unit>
  <unit id="v4">
   <val:validation>
     <val:rule isPresent="loja" occurs="2" />
   </val:validation>
   <segment id="1">
     <source>Choose a store option in the online store.</source>
     <target>Escolha uma opção de loja na online store.</target>
   </segment>
  </unit>
  <unit id="v5">
   <val:validation>
     <val:rule isNotPresent="store" />
   </val:validation>
   <segment id="1">
     <source>Choose an option in the online store.</source>
     <target>Escolha uma opção na loja online.</target>
   </segment>
  </unit>
  <unit id="v6">
   <val:validation>
     <val:rule isNotPresent="store" />
   </val:validation>
   <segment id="1">
     <source>Choose an option in the online store.</source>
     <target>Escolha uma opção na online store.</target>
   </segment>
  </unit>
  <unit id="v7">
```

```
  <val:validation>
    <val:rule startsWith="*" />
  </val:validation>
  <segment id="1">
    <source>*Choose an option in the online store.</source>
    <target>*Escolha uma opção na loja online.</target>
  </segment>
</unit>
<unit id="v8">
  <val:validation>
    <val:rule startsWith="*" />
  </val:validation>
  <segment id="1">
    <source>*Choose an option in the online store.</source>
    <target>Escolha uma opção na loja online.</target>
  </segment>
</unit>
<unit id="v9">
  <val:validation>
    <val:rule endsWith=":" />
  </val:validation>
  <segment id="1">
    <source>Choose an option in the online store:</source>
    <target>Escolha uma opção na loja online:</target>
  </segment>
</unit>
<unit id="v10">
  <val:validation>
    <val:rule endsWith=":" />
  </val:validation>
  <segment id="1">
    <source>Choose an option in the online store:</source>
    <target>Escolha uma opção na online: store.</target>
  </segment>
</unit>
<unit id="v11">
  <val:validation>
    <val:rule endsWith=":" existsInSource="yes" />
  </val:validation>
  <segment id="1">
    <source>Choose an option in the online store:</source>
    <target>Escolha uma opção na loja online:</target>
  </segment>
</unit>
<unit id="v12">
  <val:validation>
    <val:rule endsWith=":" existsInSource="no" />
  </val:validation>
  <segment id="1">
    <source>Choose an option in the online store.</source>
```

```
        <target>Escolha uma opção na loja online:</target>
    </unit>
    <unit id="v13">
      <val:validation>
        <val:rule endsWith=":" existsInSource="yes" />
      </val:validation>
      <segment id="1">
        <source>Choose an option in the online store.</source>
        <target>Escolha uma opção na loja online:</target>
      </segment>
    </unit>
</file>
```

Example 23.2 shows the results of validating Example 23.1. The results are captured in an
<xmrk:result> element inside each <unit> element.

Example 23.2 – Validation module: XLIFF file with validation results marked

```
<file xmlns="urn:oasis:names:tc:xliff:document:2.0" id="my1">
    <unit id="1">
      <xmrk:result/>
      <segment id="1">
        <source>Choose an option in the online store:</source>
        <target>Escolha uma opção na loja online:</target>
      </segment>
    </unit>
    <unit id="2">
      <xmrk:result>
        <disabled>yes</disabled>
        <good-isPresent>store</good-isPresent>
      </xmrk:result>
      <val:validation>
        <val:rule disabled="yes" isPresent="store"/>
      </val:validation>
      <segment id="1">
        <source>Choose an option in the application store:</source>
        <target>Escolha uma opção na application store:</target>
      </segment>
    </unit>
    <unit id="v1">
      <xmrk:result>
        <good-isPresent>online</good-isPresent>
      </xmrk:result>
      <val:validation>
        <val:rule isPresent="online"/>
      </val:validation>
      <segment id="1">
```

```
      <source>Choose an option in the online store.</source>
      <target>Escolha uma opção na loja online.</target>
  </unit>
  <unit id="v2">
    <xmrk:result>
      <bad-isPresent>loja</bad-isPresent>
    </xmrk:result>
    <val:validation>
      <val:rule isPresent="loja"/>
    </val:validation>
    <segment id="1">
      <source>Choose an option in the online store.</source>
      <target>Escolha uma opção na online store.</target>
    </segment>
  </unit>
  <unit id="v3">
    <xmrk:result>
      <good-isPresent>loja</good-isPresent>
      <occurs>2</occurs>
      <occurs-test-match>good</occurs-test-match>
    </xmrk:result>
    <val:validation>
      <val:rule isPresent="loja" occurs="2"/>
    </val:validation>
    <segment id="1">
      <source>Choose a store option in the online store.</source>
      <target>Escolha uma opção de loja na loja online.</target>
    </segment>
  </unit>
  <unit id="v4">
    <xmrk:result>
      <good-isPresent>loja</good-isPresent>
      <occurs>2</occurs>
      <occurs-test-match>bad</occurs-test-match>
    </xmrk:result>
    <val:validation>
      <val:rule isPresent="loja" occurs="2"/>
    </val:validation>
    <segment id="1">
      <source>Choose a store option in the online store.</source>
      <target>Escolha uma opção de loja na online store.</target>
    </segment>
  </unit>
  <unit id="v5">
    <xmrk:result>
      <good-isNotPresent>store</good-isNotPresent>
    </xmrk:result>
    <val:validation>
      <val:rule isNotPresent="store"/>
```

```
    </val:validation>
    <segment id="1">
      <source>Choose an option in the online store.</source>
      <target>Escolha uma opção na loja online.</target>
    </segment>
</unit>
<unit id="v6">
  <xmrk:result>
    <bad-isNotPresent>store</bad-isNotPresent>
  </xmrk:result>
  <val:validation>
    <val:rule isNotPresent="store"/>
  </val:validation>
  <segment id="1">
    <source>Choose an option in the online store.</source>
    <target>Escolha uma opção na online store.</target>
  </segment>
</unit>
<unit id="v7">
  <xmrk:result>
    <startsWith>*good</startsWith>
  </xmrk:result>
  <val:validation>
    <val:rule startsWith="*"/>
  </val:validation>
  <segment id="1">
    <source>*Choose an option in the online store.</source>
    <target>*Escolha uma opção na loja online.</target>
  </segment>
</unit>
<unit id="v8">
  <xmrk:result>
    <startsWith>*bad</startsWith>
  </xmrk:result>
  <val:validation>
    <val:rule startsWith="*"/>
  </val:validation>
  <segment id="1">
    <source>*Choose an option in the online store.</source>
    <target>Escolha uma opção na loja online.</target>
  </segment>
</unit>
<unit id="v9">
  <xmrk:result>
    <endsWith>:good</endsWith>
  </xmrk:result>
  <val:validation>
    <val:rule endsWith=":"/>
  </val:validation>
  <segment id="1">
```

```
      <source>Choose an option in the online store:</source>
      <target>Escolha uma opção na loja online:</target>
</unit>
<unit id="v10">
  <xmrk:result>
    <endsWith>:bad</endsWith>
  </xmrk:result>
  <val:validation>
    <val:rule endsWith=":"/>
  </val:validation>
  <segment id="1">
    <source>Choose an option in the online store:</source>
    <target>Escolha uma opção na online: store.</target>
  </segment>
</unit>
<unit id="v11">
  <xmrk:result>
    <endsWith>:good</endsWith>
    <existsInSource>
      <endsWith>:good</endsWith>
    </existsInSource>
  </xmrk:result>
  <val:validation>
    <val:rule endsWith=":" existsInSource="yes"/>
  </val:validation>
  <segment id="1">
    <source>Choose an option in the online store:</source>
    <target>Escolha uma opção na loja online:</target>
  </segment>
</unit>
<unit id="v12">
  <xmrk:result>
    <endsWith>:good</endsWith>
  </xmrk:result>
  <val:validation>
    <val:rule endsWith=":" existsInSource="no"/>
  </val:validation>
  <segment id="1">
    <source>Choose an option in the online store.</source>
    <target>Escolha uma opção na loja online:</target>
  </segment>
</unit>
<unit id="v13">
  <xmrk:result>
    <endsWith>:good</endsWith>
    <existsInSource>
      <endsWith>:bad</endsWith>
    </existsInSource>
  </xmrk:result>
```

```
      <val:validation>
        <val:rule endsWith=":" existsInSource="yes"/>
      </val:validation>
      <segment id="1">
        <source>Choose an option in the online store.</source>
        <target>Escolha uma opção na loja online:</target>
      </segment>
    </unit>
</file>
```

As with the other modules, the Validation module does not define the output format for results of a validation, it only defines the markup for defining what the validation parameters are.

Appendices

XLIFF 2.0 Structure

This appendix provides an overview of the structure of the elements and attributes in the XLIFF 2.0 standard.[1] The standard describes each element and attribute and can be used as a supplement to this book. XLIFF 2.0 is not backward compatible with the previous version, XLIFF 1.2.

XLIFF elements and hierarchy

The figures in this appendix are split into three parts: the top-level syntax (Figure A.1), the structure under the `<group>` element (Figure A.2) and the structure under the `<unit>` element (Figure A.3). The placeholder element `<other>` shows where you may add an extension from a custom namespace. Table A.1 defines the symbols used to show whether an element is required or optional and whether the element may be repeated.

Table A.1 – Legend

Symbol	Meaning
!	exactly one
+	one or more
?	zero or one
*	zero or more

---

[1] http://docs.oasis-open.org/xliff/xliff-core/v2.0/xliff-core-v2.0.html

```
<xliff>
|
+---<file> +
|       |
|       +---<skeleton> ?
|       |   |
|       |   +---<other> *
|       |
|       +---Metadata module elements ?
|       |
|       +---Resource Data module elements *
|       |
|       +---Change Tracking module elements ?
|       |
|       +---Size Restriction module elements ?
|       |
|       +---Validation module elements ?
|       |
|       +---<other> *
|       |
|       +---<notes> ?
|       |   |
|       |   +---<note> +
|       |
|       +--- (<group> OR <unit>) +
```

Figure A.1 – XLIFF top-level structure

```
<group>
|
+---<other> (including XLIFF Modules) *
|
+---<notes> ?
|   |
|   +---<note> +
|
+--- (<group> OR <unit>) +
```

Figure A.2 – Structure of the `<group>` element

```
<unit>
 |
 +---<other> (including XLIFF Modules) *
 |
 +---<notes> ?
 |    |
 |    +---<note> +
 |
 +---<originalData> ?
 |    |
 |    +---<Data> +
 |
 +---<segment> +
 |    |
 |    +---<source> 1
 |    |
 |    +---<target> ?
 |
 +---<ignorable> *
      |
      +---<source> 1
      |
      +---<target> ?
```

Figure A.3 – Structure of the `<unit>` element

XLIFF core elements

`<xliff>`

An XLIFF file uses the `<xliff>` element as the root element. Each XLIFF file contains one or more `<file>` elements. It has two required attributes, version and srcLang, and one optional attribute, trgLang.

`<file>`

The `<file>` element contains the localization material from a single source, typically one file. It has one required attribute, id, and ten optional attributes, canResegment, original, srcDir, trgDir, fs:fs, fs:subFs, slr:sizeRestriction, slr:sizeInfo, slr:sizeInfoRef, and slr:storageRestriction. This element also allows attributes from custom namespaces. The `<file>` element may contain an optional `<skeleton>` element, followed by zero or more elements from XLIFF modules or extensions in any namespace, followed by zero or one notes, followed by one or more <group> or <unit> elements.

<skeleton>

The <skeleton> element contains information that represents the structure of the source document whose text is contained in the parent <file> element. The contents of the <skeleton> element are implementation specific but are typically created by extractor agents and then used by merger agents to recreate the structure of the original file. This element has one optional attribute, href, which points to an external file that contains the structural information. The href attribute is only allowed when the <skeleton> element is empty.

<notes>

The <notes> element is the container for one or more <note> elements.

<note>

The <note> element contains user-readable comments and annotations.

<group>

The <group> element allows you to structure <unit> elements into a hierarchy. It has one required attribute, id, and ten optional attributes, canResegment, original, srcDir, trgDir, fs:fs, fs:subFs, slr:sizeRestriction, slr:sizeInfo, slr:sizeInfoRef, and slr:storageRestriction. It also allows attributes from any other namespace.

<unit>

The <unit> element is where the translations and the core translation functions take place. <unit> allows optional elements from XLIFF modules or custom namespaces, followed by an optional <notes> element and an optional <originalData> element, followed by one or more elements and zero or more <ignorable> elements. It has one required attribute, id, and the following optional attributes: canResegment, original, srcDir, trgDir, fs:fs, fs:subFs, slr:sizeRestriction, slr:sizeInfo, slr:storageRestriction, and slr:sizeInfoRef. It also allows attributes from any other namespace.

**

The `<segment>` element contains the smallest unit of translatable text, typically a sentence. It contains one `<source>` element followed by an optional `<target>` element. It has four optional attributes: `id`, `canResegment`, `state`, `subState`.

<ignorable>

The `<ignorable>` element contains data that needs to be preserved but does not need to be edited or translated. It contains one `<source>` element followed by an optional `<target>` element. It has one optional attribute: `id`.

<originalData>

The `<originalData>` element is a wrapper for one or more `<data>` elements that contain original data for any inline markup or code in a `<unit>`.

<Data>

The `<Data>` element contains the original data for inline markup or code. One or more `<Data>` elements can be contained in an `<originaldata>` element.

<source>

The `<source>` element contains the original source text for the parent `<unit>`. Once the text has been translated, the translated text goes into an adjacent `<target>` element. It has two optional attributes, `xml:lang`, `xml:space`. The text may include any of the inline elements listed in the section titled "Inline elements."

<target>

The `<target>` element contains the text translated from its sibling source document. It has two optional attributes: `xml:lang`, `xml:space`. The text may include any of the inline elements listed in the section titled "Inline elements."

Inline elements

The inline elements are allowed in the `<source>` or `<target>`. They include:

`<cp>`

The `<cp>` element's sole use is to represent a Unicode character that is invalid in XML. It is always empty and requires a hex attribute to store the Unicode information.

`<ph>`

`<ph>` stores standalone code in the original format (or a reference to a `<data>` element that contains the code). It has one required attribute, id, and the following optional attributes: canCopy, canDelete, canReorder, copyOf, disp, equiv, dataRef, subFlows, subType, type, fs:fs, fs:subFs, slr:equivStorage, slr:sizeInfo, slr:sizeInfoRef.

`<pc>`

`<pc>` carries metadata for a well-formed span of text. It has one required attribute, id, and the following optional attributes: canCopy, canDelete, canOverlap, canReorder, copyOf, dispEnd, dispStart, equivEnd, equivStart, dataRefEnd, dataRefStart, subFlowsEnd, subFlowsStart, subType, type, dir, fs:fs, fs:subFs, slr:storageRestriction, slr:sizeRestriction, slr:equivStorage, slr:sizeInfo, slr:sizeInfoRef.

`<sc>`

`<sc>` marks the start of a metadata span that cannot be handled with `<pc>` because the span starts in one segment and ends in another or because using `<pc>` would not result in well-formed XML. It has one required attribute, id, and the following optional attributes: canCopy, canDelete, canOverlap, canReorder, copyOf, disp, equiv, dataRef, subFlows, subType, type, dir, fs:fs, fs:subFs, slr:storageRestriction, slr:sizeRestriction, slr:equivStorage, slr:sizeInfo, slr:sizeInfoRef.

`<ec>`

`<ec>` marks the end of a span that began with `<sc>`. It has the following optional attributes: canCopy, canDelete, canOverlap, canReorder, copyOf, disp, equiv, id, isolated, dataRef, startRef, subFlows, subType, type, fs:fs, fs:subFs, slr:equivStorage, slr:sizeInfo, slr:sizeInfoRef.

`<mrk>`

`<mrk>` provides an annotation for a span of text. It has one required attribute, `id`, and the following optional attributes: `translate`, `type`, `ref`, `value`, `fs:fs`, `fs:subFs`, `slr:storageRestriction`, `slr:sizeRestriction`, and attributes from any namespace.

`<sm>`

`<sm>` marks the start of an annotation that cannot be handled with `<mrk>` because the span starts in one segment and ends in another or because using `<mrk>` would not result in well-formed XML. It has one required attribute, `id`, and the following optional attributes: `translate`, `type`, `ref`, `value`, `fs:fs`, `fs:subFs`, `slr:storageRestriction`, `slr:sizeRestriction`, and attributes from any namespace.

`<em>`

`<em>` marks the end of an annotation that began with `<sm>`. It has one required attribute: `startRef`, which refers to the `id` attribute on the corresponding `<sm>` element.

Using XLIFF inline elements

The XLIFF inline markup elements provide a way to capture markup and sub-flows from the source text. You can use the inline markup elements in the `<source>` and `<target>` elements.

Example A.1 contains an HTML paragraph with a span of text marked up with the HTML `<b>` element. Example A.2 shows one way to capture the HTML inline markup using the XLIFF `<pc>` element. The `type` and `subType` attributes tell us that the type of markup is formatting and the markup is bold (the XLIFF standard sets aside the namespace prefix `xlf:` for formatting information and assigns values for some common formats, including bold, italics, and underline. Other values may be assigned by users).

Example A.1 – HTML with inline markup

```
<p>The swallow is <b>fast</b>.</p>
```

Example A.2 – XLIFF for Example A.2 with inline markup

```
<unit id="1">
  <segment>
    <source>
      The swallow is <pc id="p1" type="fmt" subType="xlf:b">fast</pc>.
    </source>
  </segment>
</unit>
```

To capture markup from the original source that does not contain any translatable text – for example, to capture an empty element such as `<br/>` – you can use the `<ph>` element. This element is always empty; it captures the necessary information in attributes or by pointing to a `<data>` element. Example A.3 shows an alternative way to capture the markup in Example A.1, using the `<originalData>` and `<Data>` elements. These elements contain the original markup from the source document.

Example A.3 – XLIFF for Example A.2 using the `<ph>` element

```
<unit id="1">
  <originalData>
    <data id="d1">&lt;b></data>
    <data id="d2">&lt;/b></data>
  </originalData>
  <segment>
    <source>
      The swallow is <ph id="p1" dataRef="d1"/>fast<ph id="p2" dataRef="d2"/>.
    </source>
  </segment>
</unit>
```

The `<pc>` element represents well-formed spans from the source. When a span is not well formed or if a span starts in one segment and ends in another, the `<sc>` and `<ec>` elements are used. For example, the HTML in Example A.4 contains two sentences with a `<b>` span that goes from the middle of one sentence to the middle of another. Since an example like this is normally split into two segments for translation, we need to use `<sc>` and `<ec>` to capture the markup. Example A.5 shows the resulting XLIFF file.

Example A.4 – HTML with inline markup that spans sentences

```
<p>The swallow is <b>fast. The woodcock</b> is slow.</p>
```

Example A.5 – XLIFF for Example A.4 using the `<sc>` and `<ec>` elements

```
<unit id="1">
  <segment>
    <source>
      The swallow is <sc id="b1" type="fmt" subType="xlf:b"/>fast.
    </source>
  </segment>
  <segment>
    <source>
      The woodcock<ec startRef="b1" type="fmt" subType="xlf:b"/> is slow.
    </source>
  </segment>
</unit>
```

The `<mrk>` element is used to represent an annotation. It may contain text and any other inline element in any order. `<mrk>` has a start and end tag and spans text: `<mrk>some text</mrk>`

When an annotation spans more than one segment or would result in badly formed XML, you can use the `<sm>` and `<em>` elements. The `<sm>` element marks the start of an annotation. It is always empty. The `<em>` element marks the end of an annotation. It is always empty. The `startRef` attribute on the `<em>` element refers to the corresponding `<sm>` element.

XLIFF module elements

Translation Candidates module elements

The elements defined in the Translation Candidates module are: `<matches>` and `<match>`. The `<matches>` element contains one or more `<match>` elements.

Glossary module elements

The elements defined in the Glossary module are `<glossary>`, `<glossentry>`, `<term>`, `<translation>` and `<definition>`. The `<glossary>` element contains one or more `<glossentry>` elements.

Metadata module elements

The elements defined in the Metadata module are `<metadata>`, `<metagroup>`, and `<meta>`. The `<metadata>` element contains one or more `<metagroup>` elements.

Resource Data module elements

The elements defined in the Resource Data module are `<resourceData>`, `<resourceItemRef>`, `<source>`, `<target>`, and `<reference>`. `<resourceData>` contains zero or more `<resourceItemRef>` elements, followed by zero or more `<resourceItem>` elements, which contain zero or one `<source>` elements, followed by zero or one `<target>` elements, followed by zero or more `<reference>` elements.

Change Tracking module elements

The elements defined in the Change Tracking module are `<changeTrack>`, `<revisions>`, `<revision>`, and `<item>`. `<changeTrack>` contains one or more `<revisions>` elements. `<revisions>` has one required attribute, `appliesTo`, and the following optional attributes: `ref`, `currentVersion`, and attributes from any namespace.

Size Restriction module elements

The elements defined in the Size Restriction module are `<profiles>`, `<normalization>` and `<data>`. `<profiles>` contains zero or one `<normalization>` elements followed by elements from any namespace.

Validation module elements

The elements defined in the Validation module are `<validation>` and `<rule>`. `<validation>` contains one or more `<rule>` elements. It allows optional attributes from any namespace.

Using the XLIFF 2.0 specification

In this section, we explain how to reference the XLIFF 2.0 specification. We also explain its organization and show how to use it.

Specification formats

The XLIFF 2.0 Specification is available in three formats, HTML, PDF, and XML (DocBook).

The HTML version of the specification[2] is extensively linked, and designed for easy online navigation. The PDF version of the specification[3] format is useful to download, print, and use as a local resource.

---

[2] http://docs.oasis-open.org/xliff/xliff-core/v2.0/os/xliff-core-v2.0-os.html
[3] http://docs.oasis-open.org/xliff/xliff-core/v2.0/os/xliff-core-v2.0-os.pdf

Navigating the XLIFF specification

The specification contains five main sections and three appendices.

Introduction	Defines the terminology and identifies normative and non-normative references.
Conformance	Describes how to make a conforming XLIFF 2.0 file.
Fragment Identification	Describes the fragment identification feature, including selectors and relative references.
Core Specification	Describes the general processing requirements, then defines and demonstrates the core XLIFF elements and attributes. It also describes the use of CDATA sections, XML comments, inline content, segmentation, and extension mechanisms.
Modules Specification	Describes the eight optional modules, including their elements and attributes.
Appendixes	The XLIFF 2.0 specification includes three non-normative, also known as informative, appendixes:

- XML Schemas and Catalog
- Specification Change Tracking, which shows changes made during the course of developing the standard.
- Acknowledgments

XLIFF 2.0 XML schemas

The XLIFF 2.0 XML Schemas are available on the OASIS XLIFF web site. They are required to validate XLIFF files. XLIFF 2.0 has a core schema and a separate schema for each of the optional modules.

XLIFF 2.0 core schema

The core schema[4] has all the required XLIFF 2.0 features. It uses the namespace `urn:oasis:names:tc:xliff:document:2.0`.

---

[4] http://docs.oasis-open.org/xliff/xliff-core/v2.0/os/schemas/xliff_core_2.0.xsd

XLIFF 2.0 module schemas

Each of the XLIFF modules has a schema and namespace. All of the schemas can be found at this URL: http://docs.oasis-open.org/xliff/xliff-core/v2.0/os/schemas/modules. Table A.2 shows the module name, the schema file name, and the namespace for each module.

Table A.2 – XLIFF 2.0 module schemas

Module	Schema file	Namespace
Matches	matches.xsd	urn:oasis:names:tc:xliff:matches:2.0
Glossary	glossary.xsd	urn:oasis:names:tc:xliff:glossary:2.0
Format Style	fs.xsd	urn:oasis:names:tc:xliff:fs:2.0
Metadata	metadata.xsd	urn:oasis:names:tc:xliff:metadata:2.0
Resource Data	resource_data.xsd	urn:oasis:names:tc:xliff:resourcedata:2.0
Change Tracking	change_tracking.xsd	urn:oasis:names:tc:xliff:changetracking:2.0
Size Restriction	size_restriction.xsd	urn:oasis:names:tc:xliff:sizerestriction:2.0
Validation	validation.xsd	urn:oasis:names:tc:xliff:validation:2.0

Declaring the schema in an XLIFF document

XLIFF files must declare the XLIFF Schema in the <xliff> element. In Example A.6, line 1 declares the XLIFF 2.0 namespace, and line 2 declares the W3C XMLSchema-Instance namespace (xsi). The second declaration also enables the built-in XSD attributes. Lines 3–4 point to the filename that contains the schema for the XLIFF core namespace. The xsi:schemaLocation attribute contains a sequence of namespace/URL pairs. In this instance there is just one pair for the XLIFF schema. However, you can add pairs to locate the schema for other namespaces. Some tools need the xsi namespace declaration and xsi:schemaLocation attribute, and some do not. However, unless you are sure your tool does not need these two attributes, we recommend you include them.

Example A.6 – XLIFF namespace and schema declarations

```
1  <xliff xmlns="urn:oasis:names:tc:xliff:document:2.0"
2         xmlns:xsi="http://www.w3.org/2001/XMLSchema-instance"
3         xsi:schemaLocation="urn:oasis:names:tc:xliff:document:2.0
4                             urn:xliff_core_2.0.xsd"
5         srcLang="en"
6         version="2.0">
```

XLIFF 2.0 catalog

To make XML validation and processing more efficient, the XLIFF developers have provided an XML catalog,[5] which contains information about where a parser can find local copies of the XLIFF schemas.

Most XML parsers can be used in two modes: validating and forgiving. A validating XML parser needs access to the schema(s) used in the document being parsed. The schemas are declared in the header of the XML file, but the declaration does not have to provide an exact location, such as a URI. Even if the declaration contains a URI, using that URI may be difficult or impossible if the URI points to a location on the Internet, and even if the schema(s) can be found, validation can be slowed down as the parser loads the schema(s).

XML catalogs let you store local copies of the XML schemas you use, giving you quicker access to them and letting you operate when you're not connected to the Internet. One of the biggest interoperability problems with XLIFF has been that some tools don't use XML parsers and write invalid XML files. For example, there were so many cases of files that contained invalid XML characters that the developers of XLIFF 2.0 added elements to handle characters not supported by the XML standard. Therefore, implementers need good validating parsers to ensure valid XLIFF content. The XLIFF 2.0 catalog makes it possible for you to ensure that your parser has easy access to the XML schemas it needs to work effectively.

---

[5] http://docs.oasis-open.org/xliff/xliff-core/v2.0/csprd01/schemas/catalog.xml

XSL Examples: Transforming Source to and from XLIFF

Transforming to and from web XLIFF using XSLT

Even in simple HTML websites, where most translatable text comes from HTML elements, you may need to translate text from other sources in the HTML. For example, the `alt` attribute on images and `title` attribute on links may contain text that needs to be translated. In some use cases, translatable text can occur in comments or CDATA sections. As with text from HTML elements, text from these sources must be extracted to `<unit>` and `<segment>` elements to be translated.

However, for this example, all of the text that needs to be translated comes from HTML elements. In this case, each HTML file can be captured in an individual `<file>` element in the XLIFF file. The structure of the file is captured in the `<skeleton>` element, and the translatable text is captured in `<unit>` and `<segment>` elements.

Consider the simple website shown in Figure B.1 and the six examples that follow. This is the same website we used in Chapter 5.

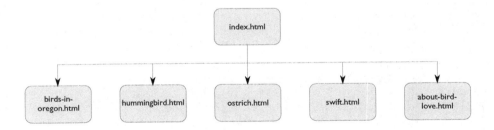

Figure B.1 – Simple HTML website

Example B.1 – HTML website: `index.html`

```html
<html>
  <head></head>
  <body>
    <h1>Bird Love</h1>
    <table>
      <tr>
        <td>
          <a href="birds-in-oregon.html">Birds in Oregon</a>
        </td>
        <td>
          <a href="hummingbird.html">Hummingbird</a>
        </td>
        <td>
          <a href="ostrich.html">Ostrich</a>
        </td>
        <td>
          <a href="swift.html">Swift</a>
        </td>
        <td>
          <a href="about-bird-love.html">About Bird Love</a>
        </td>
      </tr>
    </table>
  </body>
</html>
```

Example B.2 – HTML website: `hummingbird.html`

```html
<html id="hummingbird"
      xml:lang="en-US">
  <head></head>
  <body>
    <h2>Hummingbird</h2>
    <p>Smallest bird: Bee hummingbird (2-1/4 in)</p>
  </body>
</html>
```

Example B.3 – HTML website: `birds-in-oregon.html`

```
<html id="d1">
  <head></head>
  <body>
    <div id="s1">
      <h2>Birds in Oregon</h2>
      <p>Oregon is a mostly temperate state. There are many
      different kinds of birds that thrive there.</p>
      <div id="s1a">
        <h2>High Altitude Birds</h2>
        <p>Birds that thrive in high altitude include the
        White-tailed Ptarmigan, Sharp-tailed Grouse, Yellow-billed
        Loon, Cattle Egret, Gyrfalcon, Snowy Owl, Yellow-billed
        Cuckoo, and Boreal Owl.</p>
      </div>
      <div id="s1b">
        <h2>Ocean Birds</h2>
        <p>Common ocean birds are Brandt's Cormorant,
        Double-crested Cormorant, Pelagic Cormorant, Pigeon
        Guillemot, and Tufted Puffin.</p>
      </div>
      <div id="s1c">
        <h2>Desert Birds</h2>
        <p>Birds you find in the desert include the Sage Grouse,
        California Quail, and Prairie Falcon.</p>
      </div>
    </div>
  </body>
</html>
```

Example B.4 – HTML website: `ostrich.html`

```
<html id="ostrich"
      xml:lang="en-US">
  <head></head>
  <body>
    <h2>Ostrich</h2>
    <p>Heaviest bird: Ostrich (330 lb)</p>
  </body>
</html>
```

Example B.5 – HTML website: `swift.html`

```
<html id="swift"
      xml:lang="en-US">
  <head></head>
  <body>
    <h2>Swift</h2>
    <p>Fastest bird flying: Common Swift (125 mi/hr)</p>
  </body>
</html>
```

Example B.6 – HTML website: `about-bird-love.html`

```
<html id="abl"
      xml:lang="en-US">
  <head></head>
  <body>
    <h2>About Bird Love</h2>
    <p>Bird Love is an organization that exists for the love of
    birds.</p>
  </body>
</html>
```

Now, let's apply the XSLT in Example B.7 to the `index.html` file. This transform reads `index.html` and all of the files it links to and generates an XLIFF file for the entire site.

Example B.7 – XSLT to transform a simple website into XLIFF

```
<?xml version="1.0"?>
<xsl:transform xmlns:xsl="http://www.w3.org/1999/XSL/Transform"
               xmlns:xmrk="xmarker:skl" version="2.0">
  <xsl:output method="xml" indent="yes" encoding="utf-8" />

  <xsl:strip-space elements="*" />

  <xsl:template match="node()|@*">
    <xsl:copy>
      <xsl:apply-templates select="@*|node()" />
    </xsl:copy>
  </xsl:template>

  <xsl:template match="node()|@*" mode="body">
    <xsl:copy>
      <xsl:apply-templates select="@*|node()" mode="body" />
    </xsl:copy>
  </xsl:template>
```

```
<xsl:template match="node()|@*" mode="head">
  <xsl:copy>
    <xsl:apply-templates select="@*|node()" mode="head" />
  </xsl:copy>
</xsl:template>

<xsl:template match="html">
  <xliff version="2.0" srcLang="en" trgLang="es"
         xmlns="urn:oasis:names:tc:xliff:document:2.0"
         xmlns:xmrk="urn.xmarker.com">
    <file id="{generate-id()}" original="index.html">
      <skeleton>
        <xsl:apply-templates mode="head">
          <xsl:with-param name="myid" select="generate-id()" />
        </xsl:apply-templates>
      </skeleton>
      <xsl:apply-templates mode="body"></xsl:apply-templates>
    </file>
    <xsl:for-each select="//*[@href]">
      <xsl:variable name="file" select="@href" />
      <file original="{@href}" id="{generate-id()}">
        <skeleton>
          <xsl:apply-templates select="document($file)" mode="head"/>
        </skeleton>
        <xsl:apply-templates select="document($file)" mode="body"/>
      </file>
    </xsl:for-each>
  </xliff>
</xsl:template>

<xsl:template match="*" mode="head" priority="3">
  <xsl:variable name="myname" select="concat('xmrk:',local-name())" />
  <xsl:element name="{$myname}">
    <xsl:for-each select="@*">
      <xsl:copy />
    </xsl:for-each>
    <xsl:attribute name="idref">
      <xsl:value-of
        select="concat('d',count(preceding::*),count(ancestor::*))" />
    </xsl:attribute>
    <xsl:apply-templates mode="head" />
  </xsl:element>
</xsl:template>

<xsl:template match="text()" mode="head" priority="4">
  <xsl:text>xxxxxx</xsl:text>
</xsl:template>
```

```
<xsl:template match="*" mode="body" priority="3">
  <xsl:choose>
    <xsl:when test="text() and (preceding-sibling::text())">
      <xsl:variable name="myname"
                    select="concat('xmrk:',local-name())" />
      <pc>
        <xsl:attribute name="id">
          <xsl:value-of
          select="concat('d',count(preceding::*),count(ancestor::*))" />
        </xsl:attribute>
        <xsl:apply-templates mode="body" />
      </pc>
    </xsl:when>
    <xsl:when test="text()">
      <xsl:variable name="myname"
                    select="concat('xmrk:',local-name())" />
      <unit id="{generate-id()}" name="{concat($myname,generate-id())}"
            xmlns="urn:oasis:names:tc:xliff:document:2.0">
        <xsl:attribute name="id">
          <xsl:value-of
          select="concat('d',count(preceding::*),count(ancestor::*))"/>
        </xsl:attribute>
        <segment>
          <source>
            <xsl:apply-templates mode="body" />
          </source>
          <target>
            <xsl:apply-templates mode="body" />
          </target>
        </segment>
      </unit>
    </xsl:when>
    <xsl:otherwise>
      <xsl:apply-templates mode="body" />
    </xsl:otherwise>
  </xsl:choose>
</xsl:template>
</xsl:transform>
```

Example B.8 shows the XLIFF file generated for the website. Notice that the `<target>` elements have been supplied, but with untranslated, English text.

Example B.8 – XLIFF generated from simple website using Example B.7

```xml
<?xml version="1.0" encoding="utf-8"?>
<xliff xmlns="urn:oasis:names:tc:xliff:document:2.0"
       xmlns:xmrk="urn.xmarker.com"
       version="2.0" srcLang="en" trgLang="es">
  <file id="d1e1" original="index.html">
    <skeleton>
      <xmrk:h1 xmlns:xmrk="xmarker:skl" idref="d01">xxxxxx</xmrk:h1>
      <xmrk:table xmlns:xmrk="xmarker:skl" idref="d11">
        <xmrk:tr idref="d12">
          <xmrk:td idref="d13">
            <xmrk:a href="birds-in-oregon.html"
                    idref="d14">xxxxxx</xmrk:a>
          </xmrk:td>
          <xmrk:td idref="d33">
            <xmrk:a href="hummingbird.html"
                    idref="d34">xxxxxx</xmrk:a>
          </xmrk:td>
          <xmrk:td idref="d53">
            <xmrk:a href="ostrich.html"
                    idref="d54">xxxxxx</xmrk:a>
          </xmrk:td>
          <xmrk:td idref="d73">
            <xmrk:a href="swift.html"
                    idref="d74">xxxxxx</xmrk:a>
          </xmrk:td>
          <xmrk:td idref="d93">
            <xmrk:a href="about-bird-love.html"
                    idref="d94">xxxxxx</xmrk:a>
          </xmrk:td>
        </xmrk:tr>
      </xmrk:table>
    </skeleton>
    <unit xmlns:xmrk="xmarker:skl" id="d01" name="xmrk:h1d1e2">
      <segment>
        <source>Bird Love</source>
        <target>Bird Love</target>
      </segment>
    </unit>
    <unit xmlns:xmrk="xmarker:skl" id="d14" name="xmrk:ad1e7">
      <segment>
        <source>Birds in Oregon</source>
        <target>Birds in Oregon</target>
      </segment>
    </unit>
    <unit xmlns:xmrk="xmarker:skl" id="d34" name="xmrk:ad1e10">
      <segment>
        <source>Hummingbird</source>
        <target>Hummingbird</target>
```

```
    </unit>
    <unit xmlns:xmrk="xmarker:skl" id="d54" name="xmrk:ad1e13">
      <segment>
        <source>Ostrich</source>
        <target>Ostrich</target>
      </segment>
    </unit>
    <unit xmlns:xmrk="xmarker:skl" id="d74" name="xmrk:ad1e16">
      <segment>
        <source>Swift</source>
        <target>Swift</target>
      </segment>
    </unit>
    <unit xmlns:xmrk="xmarker:skl" id="d94" name="xmrk:ad1e19">
      <segment>
        <source>About Bird Love</source>
        <target>About Bird Love</target>
      </segment>
    </unit>
  </file>
  <file original="birds-in-oregon.html" id="d1e7">
    <skeleton>
      <xmrk:html xmlns:xmrk="xmarker:skl" id="d1" idref="d00">
        <xmrk:div id="s1" idref="d01">
          <xmrk:h2 idref="d02">xxxxxx</xmrk:h2>
          <xmrk:p idref="d12">xxxxxx</xmrk:p>
          <xmrk:div id="s1a" idref="d22">
            <xmrk:h2 idref="d23">xxxxxx</xmrk:h2>
            <xmrk:p idref="d33">xxxxxx</xmrk:p>
          </xmrk:div>
          <xmrk:div id="s1b" idref="d52">
            <xmrk:h2 idref="d53">xxxxxx</xmrk:h2>
            <xmrk:p idref="d63">xxxxxx</xmrk:p>
          </xmrk:div>
          <xmrk:div id="s1c" idref="d82">
            <xmrk:h2 idref="d83">xxxxxx</xmrk:h2>
            <xmrk:p idref="d93">xxxxxx</xmrk:p>
          </xmrk:div>
        </xmrk:div>
      </xmrk:html>
    </skeleton>
    <unit xmlns:xmrk="xmarker:skl" id="d02" name="xmrk:h2d2e3">
      <segment>
        <source>Birds in Oregon</source>
        <target>Birds in Oregon</target>
      </segment>
    </unit>
    <unit xmlns:xmrk="xmarker:skl" id="d12" name="xmrk:pd2e5">
      <segment>
```

```
    <source>Oregon is a mostly temperate state. There are many
    different kinds of birds that thrive there.</source>
    <target>Oregon is a mostly temperate state. There are many
    different kinds of birds that thrive there.</target>
</unit>
<unit xmlns:xmrk="xmarker:skl" id="d23" name="xmrk:h2d2e8">
  <segment>
    <source>High Altitude Birds</source>
    <target>High Altitude Birds</target>
  </segment>
</unit>
<unit xmlns:xmrk="xmarker:skl" id="d33" name="xmrk:pd2e10">
  <segment>
    <source>Birds that thrive in the high altitude include the
    White-tailed Ptarmigan, Sharp-tailed Grouse, Yellow-billed Loon,
    Cattle Egret, Gyrfalcon, Snowy Owl, Yellow-billed Cuckoo, and
    Boreal Owl.</source>
    <target>Birds that thrive in the high altitude include the
    White-tailed Ptarmigan, Sharp-tailed Grouse, Yellow-billed Loon,
    Cattle Egret, Gyrfalcon, Snowy Owl, Yellow-billed Cuckoo, and
    Boreal Owl.</target>
  </segment>
</unit>
<unit xmlns:xmrk="xmarker:skl" id="d53" name="xmrk:h2d2e13">
  <segment>
    <source>Ocean Birds</source>
    <target>Ocean Birds</target>
  </segment>
</unit>
<unit xmlns:xmrk="xmarker:skl" id="d63" name="xmrk:pd2e15">
  <segment>
    <source>Common ocean birds are Brandt's Cormorant, Double-crested
    Cormorant, Pelagic Cormorant, Pigeon Guillemot, and Tufted
    Puffin.</source>
    <target>Common ocean birds are Brandt's Cormorant, Double-crested
    Cormorant, Pelagic Cormorant, Pigeon Guillemot, and Tufted
    Puffin.</target>
  </segment>
</unit>
<unit xmlns:xmrk="xmarker:skl" id="d83" name="xmrk:h2d2e18">
  <segment>
    <source>Desert Birds</source>
    <target>Desert Birds</target>
  </segment>
</unit>
<unit xmlns:xmrk="xmarker:skl" id="d93" name="xmrk:pd2e20">
  <segment>
    <source>Birds you find in the desert include the Sage Grouse,
    California Quail, and Prairie Falcon.</source>
```

```
          <target>Birds you find in the desert include the Sage Grouse,
          California Quail, and Prairie Falcon.</target>
    </unit>
</file>
<file original="hummingbird.html" id="d1e10">
    <skeleton>
      <xmrk:html xmlns:xmrk="xmarker:skl" id="hummingbird"
                 xml:lang="en-US" idref="d00">
        <xmrk:h2 idref="d01">xxxxxx</xmrk:h2>
        <xmrk:p idref="d11">xxxxxx</xmrk:p>
      </xmrk:html>
    </skeleton>
    <unit xmlns:xmrk="xmarker:skl" id="d01" name="xmrk:h2d3e2">
      <segment>
        <source>Hummingbird</source>
        <target>Hummingbird</target>
      </segment>
    </unit>
    <unit xmlns:xmrk="xmarker:skl" id="d11" name="xmrk:pd3e4">
      <segment>
        <source>Smallest bird: Bee hummingbird (2-1/4 in)</source>
        <target>Smallest bird: Bee hummingbird (2-1/4 in)</target>
      </segment>
    </unit>
</file>
<file original="ostrich.html" id="d1e13">
    <skeleton>
      <xmrk:html xmlns:xmrk="xmarker:skl" id="ostrich"
                 xml:lang="en-US" idref="d00">
        <xmrk:h2 idref="d01">xxxxxx</xmrk:h2>
        <xmrk:p idref="d11">xxxxxx</xmrk:p>
      </xmrk:html>
    </skeleton>
    <unit xmlns:xmrk="xmarker:skl" id="d01" name="xmrk:h2d4e2">
      <segment>
        <source>Ostrich</source>
        <target>Ostrich</target>
      </segment>
    </unit>
    <unit xmlns:xmrk="xmarker:skl" id="d11" name="xmrk:pd4e4">
      <segment>
        <source>Heaviest bird: Ostrich (330 lb)</source>
        <target>Heaviest bird: Ostrich (330 lb)</target>
      </segment>
    </unit>
</file>
<file original="swift.html" id="d1e16">
    <skeleton>
      <xmrk:html xmlns:xmrk="xmarker:skl" id="swift"
```

```
                    xml:lang="en-US" idref="d00">
        <xmrk:h2 idref="d01">xxxxxx</xmrk:h2>
        <xmrk:p idref="d11">xxxxxx</xmrk:p>
      </xmrk:html>
    </skeleton>
    <unit xmlns:xmrk="xmarker:skl" id="d01" name="xmrk:h2d5e2">
      <segment>
        <source>Swift</source>
        <target>Swift</target>
      </segment>
    </unit>
    <unit xmlns:xmrk="xmarker:skl" id="d11" name="xmrk:pd5e4">
      <segment>
        <source>Fastest bird flying: Common Swift (125 mi/hr)</source>
        <target>Fastest bird flying: Common Swift (125 mi/hr)</target>
      </segment>
    </unit>
  </file>
  <file original="about-bird-love.html" id="d1e19">
    <skeleton>
      <xmrk:html xmlns:xmrk="xmarker:skl" id="abl"
                    xml:lang="en-US" idref="d00">
        <xmrk:h2 idref="d01">xxxxxx</xmrk:h2>
        <xmrk:p idref="d11">xxxxxx</xmrk:p>
      </xmrk:html>
    </skeleton>
    <unit xmlns:xmrk="xmarker:skl" id="d01" name="xmrk:h2d6e2">
      <segment>
        <source>About Bird Love</source>
        <target>About Bird Love</target>
      </segment>
    </unit>
    <unit xmlns:xmrk="xmarker:skl" id="d11" name="xmrk:pd6e4">
      <segment>
        <source>Bird Love is an organization that exists for the love of
        birds.</source>
        <target>Bird Love is an organization that exists for the love of
        birds.</target>
      </segment>
    </unit>
  </file>
</xliff>
```

Example B.9 shows the XLIFF file after it has been translated to Spanish.

Example B.9 – XLIFF file from Example B.8 translated into Spanish

```xml
<?xml version="1.0" encoding="utf-8"?>
<xliff xmlns="urn:oasis:names:tc:xliff:document:2.0"
       xmlns:xmrk="urn.xmarker.com"
       version="2.0" srcLang="en" trgLang="es">
  <file id="d1e1" original="index.html">
    <skeleton>
      <xmrk:h1 xmlns:xmrk="xmarker:skl" idref="d01">xxxxxx</xmrk:h1>
      <xmrk:table xmlns:xmrk="xmarker:skl" idref="d11">
        <xmrk:tr idref="d12">
          <xmrk:td idref="d13">
            <xmrk:a href="birds-in-oregon.html"
                    idref="d14">xxxxxx</xmrk:a>
          </xmrk:td>
          <xmrk:td idref="d33">
            <xmrk:a href="hummingbird.html"
                    idref="d34">xxxxxx</xmrk:a>
          </xmrk:td>
          <xmrk:td idref="d53">
            <xmrk:a href="ostrich.html"
                    idref="d54">xxxxxx</xmrk:a>
          </xmrk:td>
          <xmrk:td idref="d73">
            <xmrk:a href="swift.html"
                    idref="d74">xxxxxx</xmrk:a>
          </xmrk:td>
          <xmrk:td idref="d93">
            <xmrk:a href="about-bird-love.html"
                    idref="d94">xxxxxx</xmrk:a>
          </xmrk:td>
        </xmrk:tr>
      </xmrk:table>
    </skeleton>
    <unit xmlns:xmrk="xmarker:skl" id="d01" name="xmrk:h1d1e2">
      <segment>
        <source>Bird Love</source>
        <target>Amor por los pájaros</target>
      </segment>
    </unit>
    <unit xmlns:xmrk="xmarker:skl" id="d14" name="xmrk:ad1e7">
      <segment>
        <source>Birds in Oregon</source>
        <target>Pájaros en Oregon</target>
      </segment>
    </unit>
    <unit xmlns:xmrk="xmarker:skl" id="d34" name="xmrk:ad1e10">
      <segment>
        <source>Hummingbird</source>
        <target>Colibrí</target>
```

```
    </unit>
    <unit xmlns:xmrk="xmarker:skl" id="d54" name="xmrk:ad1e13">
      <segment>
        <source>Ostrich</source>
        <target>Avestruz</target>
      </segment>
    </unit>
    <unit xmlns:xmrk="xmarker:skl" id="d74" name="xmrk:ad1e16">
      <segment>
        <source>Swift</source>
        <target>Vencejo</target>
      </segment>
    </unit>
    <unit xmlns:xmrk="xmarker:skl" id="d94" name="xmrk:ad1e19">
      <segment>
        <source>About Bird Love</source>
        <target>Acerca de "Amor por los pájaros"</target>
      </segment>
    </unit>
  </file>
  <file original="birds-in-oregon.html" id="d1e7">
    <skeleton>
      <xmrk:html xmlns:xmrk="xmarker:skl" id="d1" idref="d00">
        <xmrk:div id="s1" idref="d01">
          <xmrk:h2 idref="d02">xxxxxx</xmrk:h2>
          <xmrk:p idref="d12">xxxxxx</xmrk:p>
          <xmrk:div id="s1a" idref="d22">
            <xmrk:h2 idref="d23">xxxxxx</xmrk:h2>
            <xmrk:p idref="d33">xxxxxx</xmrk:p>
          </xmrk:div>
          <xmrk:div id="s1b" idref="d52">
            <xmrk:h2 idref="d53">xxxxxx</xmrk:h2>
            <xmrk:p idref="d63">xxxxxx</xmrk:p>
          </xmrk:div>
          <xmrk:div id="s1c" idref="d82">
            <xmrk:h2 idref="d83">xxxxxx</xmrk:h2>
            <xmrk:p idref="d93">xxxxxx</xmrk:p>
          </xmrk:div>
        </xmrk:div>
      </xmrk:html>
    </skeleton>
    <unit xmlns:xmrk="xmarker:skl" id="d02" name="xmrk:h2d2e3">
      <segment>
        <source>Birds in Oregon</source>
        <target>Pájaros en Oregon</target>
      </segment>
    </unit>
    <unit xmlns:xmrk="xmarker:skl" id="d12" name="xmrk:pd2e5">
      <segment>
```

```
      <source>Oregon is a mostly temperate state. There are many
      different kinds of birds that thrive there.</source>
      <target>Oregon es un estado generalmente templado. Muchos tipos
      diferentes de pájaros prosperan allí.</target>
</unit>
<unit xmlns:xmrk="xmarker:skl" id="d23" name="xmrk:h2d2e8">
  <segment>
    <source>High Altitude Birds</source>
    <target>Pájaros de gran altura</target>
  </segment>
</unit>
<unit xmlns:xmrk="xmarker:skl" id="d33" name="xmrk:pd2e10">
  <segment>
    <source>Birds that thrive in the high altitude include the
    White-tailed Ptarmigan, Sharp-tailed Grouse, Yellow-billed Loon,
    Cattle Egret, Gyrfalcon, Snowy Owl, Yellow-billed Cuckoo, and
    Boreal Owl.</source>
    <target>Los pájaros que prosperan a grandes alturas incluyen a
    la perdiz nival de cola blanca, urogallo de las praderas, colimbo
    de Adams, garza boyera, halcón gerifalte, gran buho blanco,
    cuclillo piquigualdo y al mochuelo boreal.</target>
  </segment>
</unit>
<unit xmlns:xmrk="xmarker:skl" id="d53" name="xmrk:h2d2e13">
  <segment>
    <source>Ocean Birds</source>
    <target>Pájaros oceánicos</target>
  </segment>
</unit>
<unit xmlns:xmrk="xmarker:skl" id="d63" name="xmrk:pd2e15">
  <segment>
    <source>Common ocean birds are Brandt's Cormorant, Double-crested
    Cormorant, Pelagic Cormorant, Pigeon Guillemot, and Tufted
    Puffin.</source>
    <target>Los pájaros oceánicos comunes son el cormorán de
    Brandt, cormorán orejudo, cormorán pelágico, arao colombino y
    el frailecillo coletudo.</target>
  </segment>
</unit>
<unit xmlns:xmrk="xmarker:skl" id="d83" name="xmrk:h2d2e18">
  <segment>
    <source>Desert Birds</source>
    <target>Pájaros del desierto</target>
  </segment>
</unit>
<unit xmlns:xmrk="xmarker:skl" id="d93" name="xmrk:pd2e20">
  <segment>
    <source>Birds you find in the desert include the Sage Grouse,
    California Quail, and Prairie Falcon.</source>
```

```
        <target>Los pájaros que se encuentran en el desierto incluyen al
        urogallo de las artemisas, codorniz californiana y al halcón
        pálido.</target>
    </unit>
  </file>
  <file original="hummingbird.html" id="d1e10">
    <skeleton>
      <xmrk:html xmlns:xmrk="xmarker:skl" id="hummingbird"
                 xml:lang="en-US" idref="d00">
      <xmrk:h2 idref="d01">xxxxxx</xmrk:h2>
      <xmrk:p idref="d11">xxxxxx</xmrk:p>
      </xmrk:html>
    </skeleton>
    <unit xmlns:xmrk="xmarker:skl" id="d01" name="xmrk:h2d3e2">
      <segment>
        <source>Hummingbird</source>
        <target>Colibrí</target>
      </segment>
    </unit>
    <unit xmlns:xmrk="xmarker:skl" id="d11" name="xmrk:pd3e4">
      <segment>
        <source>Smallest bird: Bee hummingbird (2-1/4 in)</source>
        <target>Pájaro más pequeño: Colibrí zunzuncito (5
        cm)</target>
      </segment>
    </unit>
  </file>
  <file original="ostrich.html" id="d1e13">
    <skeleton>
      <xmrk:html xmlns:xmrk="xmarker:skl" id="ostrich"
                 xml:lang="en-US" idref="d00">
      <xmrk:h2 idref="d01">xxxxxx</xmrk:h2>
      <xmrk:p idref="d11">xxxxxx</xmrk:p>
      </xmrk:html>
    </skeleton>
    <unit xmlns:xmrk="xmarker:skl" id="d01" name="xmrk:h2d4e2">
      <segment>
        <source>Ostrich</source>
        <target>Avestruz</target>
      </segment>
    </unit>
    <unit xmlns:xmrk="xmarker:skl" id="d11" name="xmrk:pd4e4">
      <segment>
        <source>Heaviest bird: Ostrich (330 lb)</source>
        <target>Pájaro más pesado: Avestruz (150 Kg)</target>
      </segment>
    </unit>
  </file>
  <file original="swift.html" id="d1e16">
```

```
  <skeleton>
    <xmrk:html xmlns:xmrk="xmarker:skl" id="swift"
               xml:lang="en-US" idref="d00">
      <xmrk:h2 idref="d01">xxxxxx</xmrk:h2>
      <xmrk:p idref="d11">xxxxxx</xmrk:p>
    </xmrk:html>
  </skeleton>
  <unit xmlns:xmrk="xmarker:skl" id="d01" name="xmrk:h2d5e2">
    <segment>
      <source>Swift</source>
      <target>Vencejo</target>
    </segment>
  </unit>
  <unit xmlns:xmrk="xmarker:skl" id="d11" name="xmrk:pd5e4">
    <segment>
      <source>Fastest bird flying: Common Swift (125 mi/hr)</source>
      <target>Pájaro de vuelo más veloz: Vencejo común (200
      Km/h)</target>
    </segment>
  </unit>
</file>
<file original="about-bird-love.html" id="d1e19">
  <skeleton>
    <xmrk:html xmlns:xmrk="xmarker:skl" id="abl"
               xml:lang="en-US" idref="d00">
      <xmrk:h2 idref="d01">xxxxxx</xmrk:h2>
      <xmrk:p idref="d11">xxxxxx</xmrk:p>
    </xmrk:html>
  </skeleton>
  <unit xmlns:xmrk="xmarker:skl" id="d01" name="xmrk:h2d6e2">
    <segment>
      <source>About Bird Love</source>
      <target>Acerca de "Amor por los pájaros"</target>
    </segment>
  </unit>
  <unit xmlns:xmrk="xmarker:skl" id="d11" name="xmrk:pd6e4">
    <segment>
      <source>Bird Love is an organization that exists for the love of
      birds.</source>
      <target>"Amor por los pájaros" es una organización que existe
      por amor a los pájaros.</target>
    </segment>
  </unit>
</file>
</xliff>
```

The XSLT in Example B.10 merges the translation back into the original files.

Example B.10 – XSLT to transform XLIFF back into a website

```
<?xml version="1.0"?>
<xsl:transform xmlns:xsl="http://www.w3.org/1999/XSL/Transform"
               xmlns:xlf="urn:oasis:names:tc:xliff:document:2.0"
               xmlns:xmrk="xmarker:skl" version="2.0">
  <xsl:output method="xml" indent="yes" encoding="utf-8" />
  <xsl:strip-space elements="*" />
  <xsl:template match="node()|@*">
    <xsl:copy>
      <xsl:apply-templates select="@*|node()" />
    </xsl:copy>
  </xsl:template>

  <xsl:template match="node()|@*" mode="skl">
    <xsl:copy>
      <xsl:apply-templates select="@*|node()" mode="skl" />
    </xsl:copy>
  </xsl:template>

  <xsl:template match="xlf:skeleton" priority="3">
    <xsl:variable name="myname">
      <xsl:for-each select="ancestor::*[local-name()='file']">
        <xsl:value-of select="@original" />
      </xsl:for-each>
    </xsl:variable>
    <xsl:choose>
      <xsl:when test="not($myname='index.html')">
        <xsl:result-document href="{concat('es-',$myname)}">
          <xsl:apply-templates mode="skl" />
        </xsl:result-document>
      </xsl:when>
      <xsl:otherwise>
        <html id="{$myname}">
          <xsl:apply-templates mode="skl" />
        </html>
      </xsl:otherwise>
    </xsl:choose>
  </xsl:template>

  <xsl:template match="xlf:xliff|xlf:file" priority="3">
    <xsl:apply-templates />
  </xsl:template>

  <xsl:template match="*[ancestor::xlf:skeleton]" priority="2" mode="skl">
    <xsl:variable name="idref" select="@idref" />
    <xsl:variable name="file">
      <xsl:for-each select="ancestor::xlf:file">
        <xsl:value-of select="count(preceding-sibling::*)" />
      </xsl:for-each>
```

```
      </xsl:variable>
      <xsl:element name="{local-name()}">
        <xsl:for-each
         select="@*[not(local-name()='href')and not(local-name()='idref')]">
          <xsl:copy />
        </xsl:for-each>
        <xsl:for-each select="@href">
          <xsl:attribute name="href">
            <xsl:value-of select="concat('es-',.)" />
          </xsl:attribute>
        </xsl:for-each>
        <xsl:choose>
          <xsl:when test="* and not(text())">
            <xsl:apply-templates mode="skl" />
          </xsl:when>
          <xsl:when test="* and text()">
            <xsl:for-each select="ancestor::xlf:file">
              <xsl:apply-templates
                  select=".//xlf:unit[@id=$idref]/xlf:segment/xlf:target"
                  mode="skl" />
            </xsl:for-each>
          </xsl:when>
          <xsl:when test="text()">
            <xsl:for-each select="ancestor::xlf:file">
              <xsl:apply-templates
                  select=".//xlf:unit[@id=$idref]/xlf:segment/xlf:target"
                  mode="skl" />
            </xsl:for-each>
          </xsl:when>
          <xsl:otherwise>
            <xsl:apply-templates mode="skl" />
          </xsl:otherwise>
        </xsl:choose>
      </xsl:element>
    </xsl:template>
    <xsl:template match="xlf:target" priority="3" mode="skl">
      <xsl:apply-templates mode="skl" />
    </xsl:template>
    <xsl:template match="xsl:mrk" priority="3" mode="skl">
      <xsl:variable name="myid" select="@id" />
      <xsl:variable name="myname">
        <xsl:for-each select="ancestor::xlf:file//*[@idref=$myid]">
          <xsl:value-of select="local-name()" />
        </xsl:for-each>
      </xsl:variable>
      <xsl:element name="{$myname}">
        <xsl:apply-templates mode="skl" />
      </xsl:element>
    </xsl:template>
  </xsl:transform>
```

The result, which you can see in Chapter 5, the section titled "Translating simple HTML websites" (p. 31), is an exact copy of the original HTML in the translated language.

 You can download both of these XSL scripts from the companion website for this book: http://xmlpress.net/xliff/examples.

Transforming to and from maximalist XLIFF

The maximalist method of XLIFF processing does not use a skeleton. Instead, you capture the structure of the document using `<group>` and `<unit>` elements. Use the maximalist method when the format of the source is predictable and not overly complex. For this method to work well, you must capture enough information about the structure, element names, attribute names, and content to recreate the original file structure. This example uses the generic XML shown in Example B.11.

Example B.11 – XML source example for maximalist method: `birds.xml`

```
<document status="draft" id="d1">
 <section id="s1">
  <title>Birds in Oregon</title>
  <paragraph>Oregon is a mostly temperate state. There are many different
    kinds of birds that thrive</paragraph>
  <section id="s1a">
   <title>High Altitude Birds</title>
   <paragraph>Birds that thrive in the high altitude include the
     White-tailed Ptarmigan, Sharp-tailed Grouse, Yellow-billed Loon,
     Cattle Egret, Gyrfalcon, Snowy Owl, Yellow-billed Cuckoo, and
     Boreal Owl.</paragraph>
  </section>
  <section id="s1b">
   <title>Ocean Birds</title>
   <paragraph>Common ocean birds are Brandt's Cormorant, Double-crested
     Cormorant, Pelagic Cormorant, Pigeon Guillemot, and the Tufted Puffin.
   </paragraph>
  </section>
  <section id="s1c">
   <title>Desert Birds</title>
   <paragraph>Birds you find in the desert include the Sage Grouse,
     California Quail, and Prairie Falcon.</paragraph>
  </section>
 </section>
</document>
```

Apply the XSLT script in Example B.12 to the XML in Example B.11.

Example B.12 – XSLT to convert generic XML to maximalist XLIFF: `max-2-xlf.xsl`

```xml
<?xml version="1.0"?>
<xsl:transform xmlns:xsl="http://www.w3.org/1999/XSL/Transform"
               version="2.0">
  <xsl:output method="xml" indent="yes" encoding="utf-8" />

  <xsl:strip-space elements="*" />

  <xsl:template match="node()|@*">
    <xsl:copy>
      <xsl:apply-templates select="@*|node()" />
    </xsl:copy>
  </xsl:template>

  <xsl:template match="*[count(ancestor::*)=0]">
    <xliff xmlns="urn:oasis:names:tc:xliff:document:2.0"
           version="2.0" srcLang="en">
      <file id="{generate-id()}" original="birds.txt">
        <group id="{generate-id()}" name="{local-name()}">
          <xsl:if test="@*">
            <notes>
              <xsl:for-each select="@*">
                <xsl:variable name="name" select="name()" />
                <xsl:variable name="some2" select="." />
                <note category="{concat('attribute_',$name)}">
                  <xsl:value-of select="." />
                </note>
              </xsl:for-each>
            </notes>
          </xsl:if>
          <xsl:apply-templates />
        </group>
      </file>
    </xliff>
  </xsl:template>

  <xsl:template match="*[not(text())][count(ancestor::*)&gt;0]">
    <group xmlns="urn:oasis:names:tc:xliff:document:2.0"
           id="{generate-id()}" name="{local-name()}">
      <xsl:if test="@*">
        <notes>
          <xsl:for-each select="@*">
            <xsl:variable name="name" select="name()" />
            <xsl:variable name="some2" select="." />
            <note category="{concat('attribute_',$name)}">
              <xsl:value-of select="." />
```

```
            </note>
          </xsl:for-each>
        </notes>
      </xsl:if>
      <xsl:apply-templates />
    </group>
  </xsl:template>

  <xsl:template match="*[text()][count(ancestor::*)&gt;0]">
    <unit xmlns="urn:oasis:names:tc:xliff:document:2.0"
          id="{generate-id()}" name="{local-name()}">
      <xsl:if test="@*">
        <notes>
          <xsl:for-each select="@*">
            <xsl:variable name="name" select="name()" />
            <xsl:variable name="some2" select="." />
            <note category="{concat('attribute_',$name)}">
              <xsl:value-of select="." />
            </note>
          </xsl:for-each>
        </notes>
      </xsl:if>
      <segment>
        <source>
          <xsl:apply-templates />
        </source>
      </segment>
    </unit>
  </xsl:template>

  <xsl:template
      match="*[preceding-sibling::text() or following-sibling::text]"
      priority="3">
    <pc xmlns="urn:oasis:names:tc:xliff:document:2.0"
        id="{generate-id()}" type="other"
        subType="{concat('xmrk:',name())}">
      <xsl:apply-templates />
    </pc>
  </xsl:template>
</xsl:transform>
```

When you apply this transform to the source XML file, you get the maximalist XLIFF file shown in Example B.13.

Example B.13 – XLIFF maximalist file generated from the XSLT in Example B.12

```xml
<?xml version="1.0" encoding="utf-8"?>
<xliff xmlns="urn:oasis:names:tc:xliff:document:2.0"
       version="2.0" srcLang="en">
  <file id="d1e1" original="birds.txt">
    <group id="d1e1" name="document">
      <notes>
        <note category="attribute_status">draft</note>
        <note category="attribute_id">d1</note>
      </notes>
      <group id="d1e2" name="section">
        <notes>
          <note category="attribute_id">s1</note>
        </notes>
        <unit id="d1e3" name="title">
          <segment>
            <source>Birds in Oregon</source>
          </segment>
        </unit>
        <unit id="d1e5" name="paragraph">
          <segment>
            <source>Oregon is a mostly temperate state. There are many
            different kinds of birds that thrive</source>
          </segment>
        </unit>
        <group id="d1e7" name="section">
          <notes>
            <note category="attribute_id">s1a</note>
          </notes>
          <unit id="d1e8" name="title">
            <segment>
              <source>High Altitude Birds</source>
            </segment>
          </unit>
          <unit id="d1e10" name="paragraph">
            <segment>
              <source>Birds that thrive in the high altitude include the
              White-tailed Ptarmigan, Sharp-tailed Grouse, Yellow-billed
              Loon, Cattle Egret, Gyrfalcon, Snowy Owl, Yellow-billed
              Cuckoo, and Boreal Owl.</source>
            </segment>
          </unit>
        </group>
        <group id="d1e12" name="section">
          <notes>
            <note category="attribute_id">s1b</note>
          </notes>
          <unit id="d1e13" name="title">
            <segment>
```

```
        <source>Ocean Birds</source>
    </unit>
    <unit id="d1e15" name="paragraph">
      <segment>
        <source>Common ocean birds are Brandt's Cormorant,
        Double-crested Cormorant, Pelagic Cormorant, Pigeon
        Guillemot, and the Tufted Puffin.</source>
      </segment>
    </unit>
  </group>
  <group id="d1e17" name="section">
    <notes>
      <note category="attribute_id">s1c</note>
    </notes>
    <unit id="d1e18" name="title">
      <segment>
        <source>Desert Birds</source>
      </segment>
    </unit>
    <unit id="d1e20" name="paragraph">
      <segment>
        <source>Birds you find in the desert include the Sage
        Grouse, California Quail, and Prairie Falcon.</source>
      </segment>
    </unit>
  </group>
    </group>
   </group>
  </file>
</xliff>
```

Translate the XLIFF to get the XLIFF file shown in Example B.14.

Example B.14 – XLIFF file from Example B.13 translated into Spanish

```xml
<?xml version="1.0" encoding="utf-8"?>
<xliff xmlns="urn:oasis:names:tc:xliff:document:2.0"
       version="2.0" srcLang="en" trgLang="es">
  <file id="d1e1" original="birds.txt">
    <group id="d1e1" name="document">
      <notes>
        <note category="attribute_status">draft</note>
        <note category="attribute_id">d1</note>
      </notes>
      <group id="d1e2" name="section">
        <notes>
          <note category="attribute_id">s1</note>
        </notes>
        <unit id="d1e3" name="title">
          <segment>
            <source>Birds in Oregon</source>
            <target>Pájaros en Oregon</target>
          </segment>
        </unit>
        <unit id="d1e5" name="paragraph">
          <segment>
            <source>Oregon is a mostly temperate state. There are many
            different kinds of birds that thrive</source>
            <target>Oregon es un estado generalmente templado. Muchos
            tipos diferentes de pájaros prosperan allí.</target>
          </segment>
        </unit>
        <group id="d1e7" name="section">
          <notes>
            <note category="attribute_id">s1a</note>
          </notes>
          <unit id="d1e8" name="title">
            <segment>
              <source>High Altitude Birds</source>
              <target>Pájaros de gran altura</target>
            </segment>
          </unit>
          <unit id="d1e10" name="paragraph">
            <segment>
              <source>Birds that thrive in the high altitude include the
              White-tailed Ptarmigan, Sharp-tailed Grouse, Yellow-billed
              Loon, Cattle Egret, Gyrfalcon, Snowy Owl, Yellow-billed
              Cuckoo, and Boreal Owl.</source>
              <target>Los pájaros que prosperan a grandes alturas
              incluyen a la perdiz nival de cola blanca, urogallo de las
              praderas, colimbo de Adams, garza boyera, halcón
              gerifalte, gran buho blanco, cuclillo piquigualdo y al
              mochuelo boreal.</target>
```

```
            </unit>
          </group>
          <group id="d1e12" name="section">
            <notes>
              <note category="attribute_id">s1b</note>
            </notes>
            <unit id="d1e13" name="title">
              <segment>
                <source>Ocean Birds</source>
                <target>Pájaros oceánicos</target>
              </segment>
            </unit>
            <unit id="d1e15" name="paragraph">
              <segment>
                <source>Common ocean birds are Brandt's Cormorant,
                Double-crested Cormorant, Pelagic Cormorant, Pigeon
                Guillemot, and the Tufted Puffin.</source>
                <target>Los pájaros oceánicos comunes son el cormorán de
                Brandt, cormorán orejudo, cormorán pelágico, arao
                colombino y el frailecillo coletudo.</target>
              </segment>
            </unit>
          </group>
          <group id="d1e17" name="section">
            <notes>
              <note category="attribute_id">s1c</note>
            </notes>
            <unit id="d1e18" name="title">
              <segment>
                <source>Desert Birds</source>
                <target>Pájaros del desierto</target>
              </segment>
            </unit>
            <unit id="d1e20" name="paragraph">
              <segment>
                <source>Birds you find in the desert include the Sage
                Grouse, California Quail, and Prairie Falcon.</source>
                <target>Los pájaros que se encuentran en el desierto
                incluyen al urogallo de las artemisas, codorniz
                californiana y al halcón pálido.</target>
              </segment>
            </unit>
          </group>
        </group>
      </group>
    </file>
  </xliff>
```

Apply the XSLT script in Example B.15 to the translated XLIFF file.

Example B.15 – XSLT to convert maximalist XLIFF to XML

```
<?xml version="1.0"?>
<xsl:transform xmlns:xsl="http://www.w3.org/1999/XSL/Transform"
               xmlns:xlf="urn:oasis:names:tc:xliff:document:2.0"
               version="2.0">
  <xsl:output method="xml" indent="yes" encoding="utf-8" />
  <xsl:strip-space elements="*" />
  <xsl:template match="node()|@*">
    <xsl:copy>
      <xsl:apply-templates select="@*|node()" />
    </xsl:copy>
  </xsl:template>
  <xsl:template match="xlf:group">
    <xsl:element name="{@name}">
      <xsl:for-each select="xlf:notes/xlf:note">
        <xsl:variable name="at-name"
                      select="substring-after(@category,'_')" />
        <xsl:variable name="at-val" select="." />
        <xsl:attribute name="{$at-name}">
          <xsl:value-of select="$at-val" />
        </xsl:attribute>
      </xsl:for-each>
      <xsl:apply-templates />
    </xsl:element>
  </xsl:template>
  <xsl:template match="xlf:unit">
    <xsl:element name="{@name}">
      <xsl:apply-templates />
    </xsl:element>
  </xsl:template>
  <xsl:template
      match="*[not(local-name()='unit') and not(local-name()='group')]">

    <xsl:apply-templates />
  </xsl:template>
  <xsl:template match="xlf:notes" priority="3" />
</xsl:transform>
```

The translated result is shown in Example B.16.

 You can download any of the XSL scripts described in this chapter from the companion website for this book: http://xmlpress.net/xliff/examples.

Example B.16 – XML file from Example B.11 translated into Spanish

```
<document status="draft" id="d1">
  <section id="s1">
    <title>Pájaros en Oregon</title>
    <paragraph>Oregon es un estado generalmente templado. Muchos tipos
    diferentes de pájaros prosperan allí.</paragraph>
    <section id="s1a">
      <title>Pájaros de gran altura</title>
      <paragraph>Los pájaros que prosperan a grandes alturas incluyen a
      la perdiz nival de cola blanca, urogallo de las praderas, colimbo
      de Adams, garza boyera, halcón gerifalte, gran buho blanco,
      cuclillo piquigualdo y al mochuelo boreal.</paragraph>
    </section>
    <section id="s1b">
      <title>Pájaros oceánicos</title>
      <paragraph>Los pájaros oceánicos comunes son el cormorán de
      Brandt, cormorán orejudo, cormorán pelágico, arao colombino y el
      frailecillo coletudo.</paragraph>
    </section>
    <section id="s1c">
      <title>Pájaros del desierto</title>
      <paragraph>Los pájaros que se encuentran en el desierto incluyen
      al urogallo de las artemisas, codorniz californiana y al halcón
      pálido.</paragraph>
    </section>
  </section>
</document>
```

Transforming SVG to and from XLIFF using XSLT

Consider the SVG graphic displayed in Figure B.2.

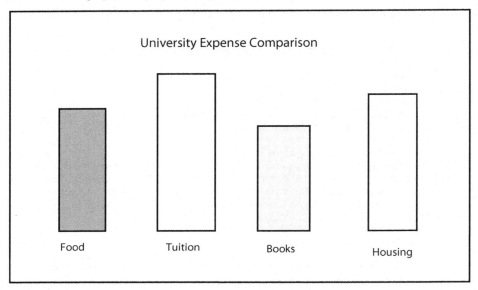

Figure B.2 – Simple SVG image

Since SVG is an XML format, we can view the XML source (see Example B.17).

Example B.17 – SVG source for Figure B.2

```
<?xml version="1.0" encoding="utf-8"?>
<svg xmlns="http://www.w3.org/2000/svg"
     width="478" height="294" viewBox="0 0 478 294"
     overflow="visible" enable-background="new 0 0 478 294"
     xml:space="preserve">
  <g id="Layer_1">
    <rect x="0.5" y="0.5" fill="#FFFFFF" stroke="#000000"
          width="477" height="293" />
    <rect x="51.5" y="105.5" fill="#99CCFF" stroke="#000000"
          width="48" height="132" />
    <rect x="153.5" y="68.5" fill="#FFFFFF" stroke="#000000"
          width="60" height="169" />
    <rect x="257.5" y="124.5" fill="#FFFF00" stroke="#000000"
          width="54" height="113" />
    <rect x="372.5" y="90.5" fill="#FFFFFF" stroke="#000000"
          width="50" height="147" />
```

```
<switch>
  <g>
    <rect x="52.5" y="250.5" fill="none" width="49"
          height="20" />
    <text id="XMLID_1_" transform="matrix(1 0 0 1 52.5 259.0195)">
      <tspan x="0" y="0" font-family="'Myriad'"
             font-size="12">Food</tspan>
    </text>
  </g>
</switch>
<switch>
  <text id="XMLID_2_" transform="matrix(1 0 0 1 162.5 259.5)">
    <tspan x="0" y="0" font-family="'Myriad'"
           font-size="12">Tuition</tspan>
  </text>
</switch>
<switch>
  <text id="XMLID_3_" transform="matrix(1 0 0 1 266.5 261.5)">
    <tspan x="0" y="0" font-family="'Myriad'"
           font-size="12">Books</tspan>
  </text>
</switch>
<switch>
  <text id="XMLID_4_" transform="matrix(1 0 0 1 376.5 265.5)">
    <tspan x="0" y="0" font-family="'Myriad'"
           font-size="12">Housing</tspan>
  </text>
</switch>
<switch>
  <text id="XMLID_5_" transform="matrix(1 0 0 1 135.5 39.5)">
    <tspan x="0" y="0" font-family="'Myriad'"
           font-size="14">University Expense Comparison</tspan>
  </text>
</switch>
  </g>
</svg>
```

We will generate an XLIFF file and an external skeleton file using the XSLT shown in Example B.18. This XSL transform creates an XLIFF file (Example B.19) containing the English text and a skeleton file, which is named `svg-skeleton-d1.txt` (Example B.20).

Example B.18 – XSLT to generate XLIFF and skeleton files

```
<?xml version="1.0"?>
<xsl:transform xmlns:xsl="http://www.w3.org/1999/XSL/Transform"
               xmlns:svg="http://www.w3.org/2000/svg" version="2.0">
  <xsl:output method="xml" indent="yes" encoding="utf-8" />
  <xsl:strip-space elements="*" />
  <xsl:variable name="file"
                select="concat('svg-skeleton-',generate-id(),'.txt')" />
  <xsl:template match="node()|@*">
    <xsl:copy>
      <xsl:apply-templates select="@*|node()" />
    </xsl:copy>
  </xsl:template>
  <xsl:template match="node()|@*" mode="skl">
    <xsl:copy>
      <xsl:apply-templates select="@*|node()" mode="skl" />
    </xsl:copy>
  </xsl:template>
  <xsl:template match="*[local-name()='svg']">
    <xsl:result-document href="{$file}">
      <svg xmlns="http://www.w3.org/2000/svg">
        <xsl:apply-templates select="node()|@*" mode="skl" />
      </svg>
    </xsl:result-document>
    <xliff xmlns="urn:oasis:names:tc:xliff:document:2.0"
           version="2.0" srcLang="en">
      <file original="my.svg" id="{generate-id()}">
        <skeleton href="{$file}" />
        <xsl:for-each select="//svg:tspan">
          <unit id="{count(preceding::*[local-name()='tspan'])}">
            <segment>
              <source>
                <xsl:apply-templates />
              </source>
            </segment>
          </unit>
        </xsl:for-each>
      </file>
    </xliff>
  </xsl:template>
  <xsl:template match="svg:tspan" mode="skl">
    <tspan xmlns="http://www.w3.org/2000/svg"
           idref="{count(preceding::*[local-name()='tspan'])}">
      <xsl:text>%%%</xsl:text>
    </tspan>
  </xsl:template>
</xsl:transform>
```

Example B.19 – XLIFF generated for the SVG file in Example B.17

```
1 <?xml version="1.0" encoding="utf-8"?>
2 <xliff xmlns="urn:oasis:names:tc:xliff:document:2.0"
3        xmlns:svg="http://www.w3.org/2000/svg"
4        version="2.0"
5        srcLang="en">
6    <file original="my.svg" id="d1e1">
7        <skeleton href="svg-skeleton-d1.txt"/>
8        <unit id="0">
9           <segment>
10              <source>Food</source>
11          </segment>
12       </unit>
13       <unit id="1">
14          <segment>
15              <source>Tuition</source>
16          </segment>
17       </unit>
18       <unit id="2">
19          <segment>
20              <source>Books</source>
21          </segment>
22       </unit>
23       <unit id="3">
24          <segment>
25              <source>Housing</source>
26          </segment>
27       </unit>
28       <unit id="4">
29          <segment>
30              <source>University Expense Comparison</source>
31          </segment>
32       </unit>
33    </file>
34 </xliff>
```

The `<skeleton href="svg-skeleton-d1.txt"/>` element on line 7 in Example B.19 refers to the external skeleton file shown in Example B.20. Since we are not embedding the skeleton in the XLIFF file, the skeleton is simply the original file with the translatable text replaced by the string "%%%." We don't need to use the `xmrk` namespace in this case.

Example B.20 – Skeleton file for Example B.17

```xml
<?xml version="1.0" encoding="utf-8"?>
<svg xmlns="http://www.w3.org/2000/svg"
    xmlns:svg="http://www.w3.org/2000/svg"
    width="478" height="294" viewBox="0 0 478 294"
    overflow="visible" enable-background="new 0 0 478 294"
    xml:space="preserve">
  <g id="Layer_1">
    <rect x="0.5" y="0.5" fill="#FFFFFF" stroke="#000000"
          width="477" height="293"/>
    <rect x="51.5" y="105.5" fill="#99CCFF" stroke="#000000"
          width="48" height="132"/>
    <rect x="153.5" y="68.5" fill="#FFFFFF" stroke="#000000"
          width="60" height="169"/>
    <rect x="257.5" y="124.5" fill="#FFFF00" stroke="#000000"
          width="54" height="113"/>
    <rect x="372.5" y="90.5" fill="#FFFFFF" stroke="#000000"
          width="50" height="147"/>
    <switch>
      <g>
        <rect x="52.5" y="250.5" fill="none" width="49" height="20"/>
        <text id="XMLID_1_" transform="matrix(1 0 0 1 52.5 259.0195)">
          <tspan idref="0" x="0" y="0"
                font-family="'Myriad'" font-size="12">%%%</tspan>
        </text>
      </g>
    </switch>
    <switch>
      <text id="XMLID_2_" transform="matrix(1 0 0 1 162.5 259.5)">
        <tspan idref="1" x="0" y="0"
                font-family="'Myriad'" font-size="12">%%%</tspan>
      </text>
    </switch>
    <switch>
      <text id="XMLID_3_" transform="matrix(1 0 0 1 266.5 261.5)">
        <tspan idref="2" x="0" y="0"
                font-family="'Myriad'" font-size="12">%%%</tspan>
      </text>
    </switch>
    <switch>
      <text id="XMLID_4_" transform="matrix(1 0 0 1 376.5 265.5)">
        <tspan idref="3" x="0" y="0"
                font-family="'Myriad'" font-size="12">%%%</tspan>
      </text>
    </switch>
    <switch>
      <text id="XMLID_5_" transform="matrix(1 0 0 1 135.5 39.5)">
        <tspan idref="4" x="0" y="0"
                font-family="'Myriad'" font-size="14">%%%</tspan>
```

```
      </text>
    </switch>
  </g>
</svg>
```

We translate the XLIFF file to German, yielding Example B.21,

Example B.21 – Translated XLIFF file for SVG example

```
<?xml version="1.0" encoding="utf-8"?>
<xliff xmlns="urn:oasis:names:tc:xliff:document:2.0"
       xmlns:svg="http://www.w3.org/2000/svg"
       version="2.0" srcLang="en" trgLang="de">
  <file original="my.svg" id="d1e1">
    <skeleton href="svg-skeleton-d1.txt" />
    <unit id="0">
      <segment>
        <source>Food</source>
        <target>Lebensmittel</target>
      </segment>
    </unit>
    <unit id="1">
      <segment>
        <source>Tuition</source>
        <target>Unterricht</target>
      </segment>
    </unit>
    <unit id="2">
      <segment>
        <source>Books</source>
        <target>Bücher</target>
      </segment>
    </unit>
    <unit id="3">
      <segment>
        <source>Housing</source>
        <target>Unterbringung</target>
      </segment>
    </unit>
    <unit id="4">
      <segment>
        <source>University Expense Comparison</source>
        <target>Universität Expense Vergleich</target>
      </segment>
    </unit>
  </file>
</xliff>
```

We then apply the XSLT shown in Example B.22, which will recreate the original file with the translated text from the skeleton file and the translated XLIFF file.

Example B.22 – XSLT to convert XLIFF file from Example B.21 back to SVG

```
<?xml version="1.0"?>
<xsl:transform xmlns:xsl="http://www.w3.org/1999/XSL/Transform"
               xmlns:svg="http://www.w3.org/2000/svg"
               xmlns:xlf="urn:oasis:names:tc:xliff:document:2.0"
               version="2.0">
  <xsl:output method="xml" indent="yes" encoding="utf-8" />
  <xsl:strip-space elements="*" />
  <xsl:variable name="file"
                select="concat('svg-skeleton-',generate-id(),'.txt')" />
  <xsl:template match="node()|@*">
    <xsl:copy>
      <xsl:apply-templates select="@*|node()" />
    </xsl:copy>
  </xsl:template>
  <xsl:template match="svg:tspan">
    <xsl:variable name="idref" select="@idref" />
    <tspan xmlns="http://www.w3.org/2000/svg">
      <xsl:for-each select="document('svg.xlf')//*[@id=$idref]">
        <xsl:value-of select="xlf:segment/xlf:target" />
      </xsl:for-each>
    </tspan>
  </xsl:template>
</xsl:transform>
```

Example B.23 shows the resulting, translated SVG file, and Figure B.3 displays the translated graphic.

Example B.23 – Translated SVG file source

```
<?xml version="1.0" encoding="utf-8"?>
<svg xmlns="http://www.w3.org/2000/svg"
     xmlns:svg="http://www.w3.org/2000/svg"
     width="478" height="294" viewBox="0 0 478 294"
     overflow="visible" enable-background="new 0 0 478 294"
     xml:space="preserve">
  <g id="Layer_1">
    <rect x="0.5" y="0.5" fill="#FFFFFF" stroke="#000000"
          width="477" height="293"/>
    <rect x="51.5" y="105.5" fill="#99CCFF" stroke="#000000"
          width="48" height="132"/>
    <rect x="153.5" y="68.5" fill="#FFFFFF" stroke="#000000"
          width="60" height="169"/>
```

```
<rect x="257.5" y="124.5" fill="#FFFF00" stroke="#000000"
      width="54" height="113"/>
<rect x="372.5" y="90.5" fill="#FFFFFF" stroke="#000000"
      width="50" height="147"/>
<switch>
  <g>
    <rect x="52.5" y="250.5" fill="none" width="49" height="20"/>
    <text id="XMLID_1_" transform="matrix(1 0 0 1 52.5 259.0195)">
      <tspan idref="0" x="0" y="0" font-family="'Myriad'"
             font-size="12">Lebensmittel</tspan>
    </text>
  </g>
</switch>
<switch>
  <text id="XMLID_2_" transform="matrix(1 0 0 1 162.5 259.5)">
    <tspan idref="1" x="0" y="0"
           font-family="'Myriad'" font-size="12">Unterricht</tspan>
  </text>
</switch>
<switch>
  <text id="XMLID_3_" transform="matrix(1 0 0 1 266.5 261.5)">
    <tspan idref="2" x="0" y="0"
           font-family="'Myriad'" font-size="12">Bücher</tspan>
  </text>
</switch>
<switch>
  <text id="XMLID_4_" transform="matrix(1 0 0 1 376.5 265.5)">
    <tspan idref="3" x="0" y="0" font-family="'Myriad'"
           font-size="12">Unterbringung</tspan>
  </text>
</switch>
<switch>
  <text id="XMLID_5_" transform="matrix(1 0 0 1 135.5 39.5)">
    <tspan idref="4" x="0" y="0" font-family="'Myriad'"
           font-size="14">Universität Expense Vergleich</tspan>
  </text>
</switch>
  </g>
</svg>
```

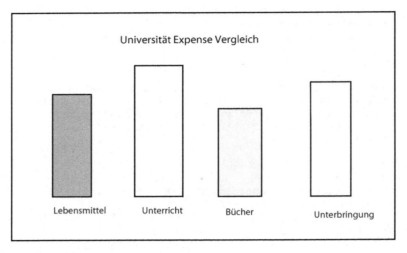

Figure B.3 – Simple SVG image in German

Glossary

character set

A collection of characters (sometimes referred to as a code page) that is associated with a sequence of natural numbers in order to facilitate the storage of text in computers and the transmission of text through telecommunication networks.

Computer Aided Translation (CAT)

A software application designed to assist human translators in the translation process.

CSV

CSV (Comma Separated Values) is a file format used to store tabular data. Each line in the file is a row in the table, and the contents of each column are separated by a delimiter, typically a comma.

example-based machine translation

A technology capable of building translations by combining data extracted from translation memories and terminology databases.

GlossML

Glossary Markup Language (GlossML)[1] is an XML vocabulary designed for containing glossaries used in translation/localization industry.

Localization Service Provider (LSP)

A company or individual that provides translation and localization services.

machine translation

A technology that automatically translates text from one language to another using previously defined grammar rules, glossaries, statistical analysis, and other methods.

OASIS

OASIS (Organization for the Advancement of Structured Information Standards)[2] is a not-for-profit consortium that drives the development, convergence, and adoption of open standards for the global information society.

---

[1] http://www.maxprograms.com/glossml/glossml.pdf
[2] http://www.oasis-open.org/

regular expression

A formula or expression that describes text strings using a specially defined syntax. For more information, see https://en.wikipedia.org/wiki/Regular_expression

skeleton

A representation of the structure of a document. In XLIFF, the `<skeleton>` element either contains the skeleton or points to an external file that contains the skeleton. The exact contents of a skeleton are implementation specific and not defined by the XLIFF standard.

source file

The original file that contains content to be translated.

source language

The language of a document that is to be translated.

SRX

Segmentation Rules eXchange (SRX) is an XML-based open standard, published by LISA (Localization Industry Standards Association),[3] for describing how translation and other language-processing tools segment text for processing.

target file

The localized output file with translated content.

target language

The language into which a document is being translated.

TBX

TBX (TermBase eXchange) is an open, XML-based standard for exchanging structured terminology data. First released by LISA in May, 2002, TBX was submitted to the International Organization for Standardization (ISO) on February 21, 2007, for adoption as an ISO standard.

TMX

Translation Memory eXchange (TMX) is an open standard originally published by LISA (Localization Industry Standards Association). The purpose of TMX is to allow easier exchange of translation memory data between tools and/or translation vendors with little or no loss of critical data during the process.

---

[3] http://www.lisa.org

translation memory

Translation Memory (TM) is a language technology that enables the translation of segments (paragraphs, sentences or phrases) of documents by searching for similar segments in a database and suggesting matches that are found in the databases as possible translations.

XLIFF

XLIFF (XML Localisation Interchange File Format) is an open standard developed by OASIS (Organization for the Advancement of Structured Information Standards).[4] The purpose of this vocabulary is to store localizable data and carry it from one step of the localization process to the other, while allowing interoperability between tools.

---

[4] http://www.oasis-open.org/

Index

CPSIA information can be obtained
at www.ICGtesting.com
Printed in the USA
LVOW03s1835021115

460758LV00023B/1196/P